Reteaching Workbook

TAKE ANOTHER LOOK

TEACHER'S EDITION
Grade 4

Harcourt Brace & Company
Orlando • Atlanta • Austin • Boston • San Francisco • Chicago • Dallas • New York • Toronto • London
http://www.hbschool.com

Printed in the United States of America

ISBN 0-15-311074-0

2 3 4 5 6 7 8 9 10 073 2000 99

CONTENTS

Name ______________________

Modeling Addition or Subtraction Situations

Using a model helps you know whether to add or subtract.

	Model	**Number Sentence**
Tina had 14 pairs of scissors. She handed out 8 pairs to classmates. How many pairs of scissors does she have now?	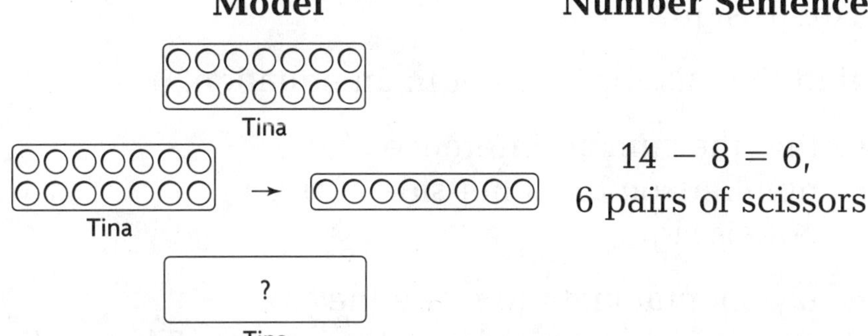	$14 - 8 = 6$, 6 pairs of scissors
Tina collected 6 pairs of scissors. Then she collected 8 more pairs. How many pairs of scissors has she collected?	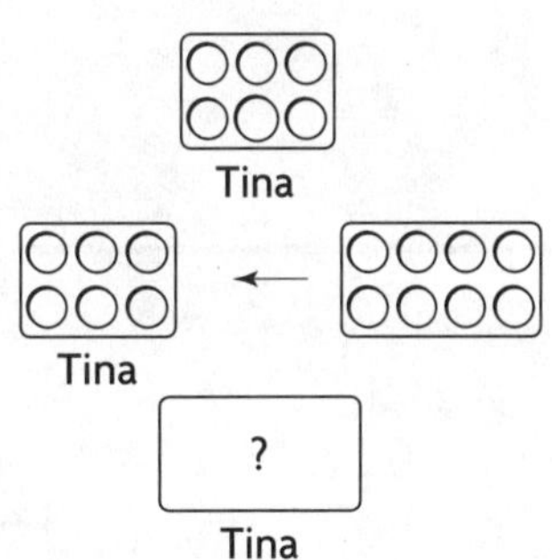	$6 + 8 = 14$, 14 pairs of scissors

Write the letter of the model that matches the problem. Then write a number sentence for each problem.

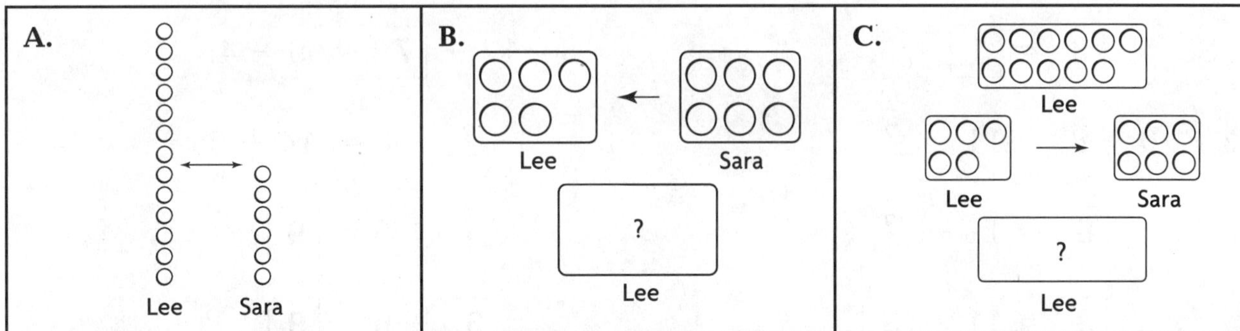

1. Lee had 5 paper clips. Sara gave him 6 more. Now how many paper clips does Lee have?

   B $5 + 6 = 11$

2. Lee has 13 paper clips. Sara has 6 paper clips. How many more paper clips does Lee have than Sara?

   A $13 - 6 = 7$

3. Lee had 11 paper clips. He gave 6 to Sara. How many paper clips does he have now?

   C $11 - 6 = 5$

Name ________________________________

LESSON 1.2

Using Mental Math

If two amounts are equal, they are the same.

An **equation** is a number sentence that uses an equals sign.

To find the missing number in an equation,

- find the sum or difference of the numbers on the other side of the equals sign.
- use mental math to make the amounts on both sides of the equals sign the same.

$3 + 7 = 2 + \underline{?}$

$10 = 2 + \underline{?}$

Think: Does $10 = 2 + 6$? No
Does $10 = 2 + 8$? Yes

So, $3 + 7 = 2 + 8$.

Complete the steps.

1. $4 + 3 = \underline{?} + 6$

$\boxed{7} = \underline{?} + 6$

$\boxed{7} = \boxed{1} + 6$

So, $4 + 3 = \boxed{1} + 6$.

2. $12 - \underline{?} = 9 - 4$

$12 - \underline{?} = \boxed{5}$

$12 - \boxed{7} = \boxed{5}$

So, $12 - \boxed{7} = 9 - 4$.

3. $0 + 8 = 11 - \underline{?}$

$\boxed{8} = 11 - \underline{?}$

$\boxed{8} = 11 - \boxed{3}$

So, $0 + 8 = 11 - \boxed{3}$.

4. $\underline{?} + 6 = 14 - 5$

$\underline{?} + 6 = \boxed{9}$

$\boxed{3} + 6 = \boxed{9}$

So, $\boxed{3} + 6 = 14 - 5$.

Use mental math to complete each equation.

5. $13 - 6 = \boxed{2} + 5$

6. $\boxed{7} + 1 = 4 + 4$

7. $12 - 8 = \boxed{13} - 9$

Name ______________________________________

Adding Three or More Addends

The **Grouping Property** states that if you group addends differently, the sum does not change.

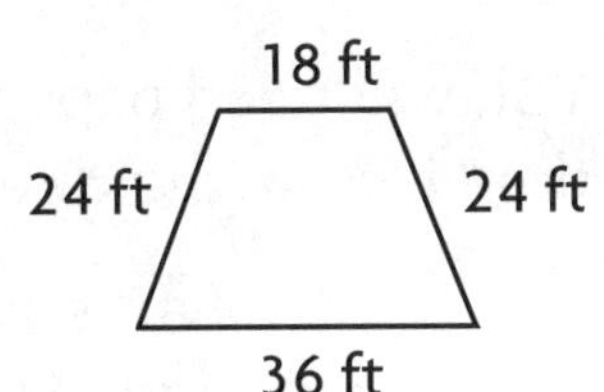

What is the perimeter of this figure?

Add to find the perimeter. 24 + 18 + 24 + 36 = __?__ ft

- Add two numbers at a time.

$$\begin{array}{r} 24 \\ +18 \\ \hline 42 \end{array} \qquad \begin{array}{r} 24 \\ +36 \\ \hline 60 \end{array}$$

- Add the sums.

$$\begin{array}{r} 42 \\ +60 \\ \hline 102 \text{ ft} \end{array}$$

So, the perimeter of this figure is 102 ft.

Find the perimeter of each figure. Then write a number sentence to show how you solved the problem. **Possible answers are given.**

1.

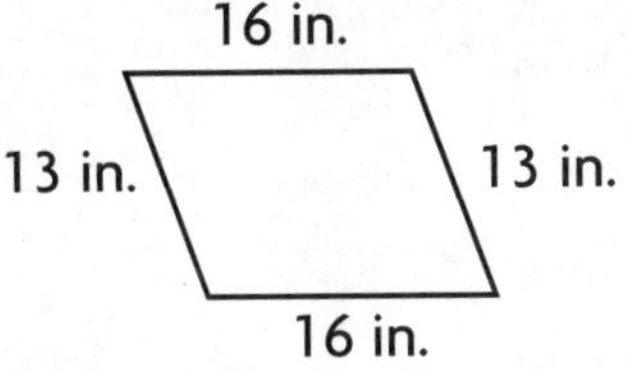

16 + 13 + 16 + 13 = __?__ in.

$$\begin{array}{r} 16 \\ +13 \\ \hline 29 \end{array} \qquad \begin{array}{r} 16 \\ +13 \\ \hline 29 \end{array} \qquad \begin{array}{r} 29 \\ +29 \\ \hline \text{58, 58 in.} \end{array}$$

2.

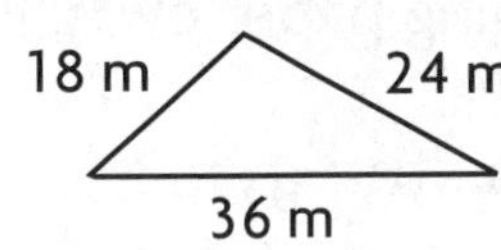

18 + 24 + 36 = __?__ m

$$\begin{array}{r} 18 \\ +24 \\ \hline 42 \end{array} \qquad \begin{array}{r} 0 \\ +36 \\ \hline 36 \end{array} \qquad \begin{array}{r} 42 \\ +36 \\ \hline \text{78, 78 m} \end{array}$$

3.

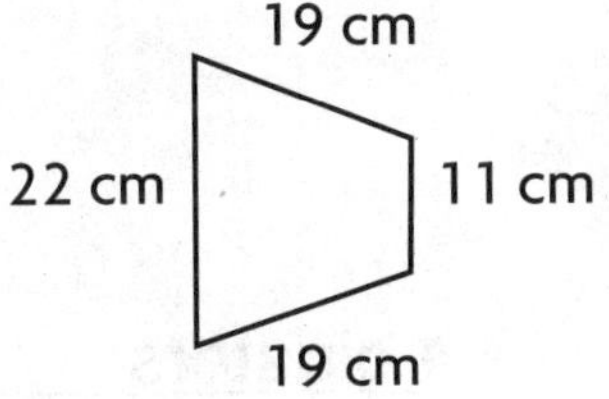

22 + 19 + 11 + 19 = __?__ cm

$$\begin{array}{r} 22 \\ +19 \\ \hline 41 \end{array} \qquad \begin{array}{r} 11 \\ +19 \\ \hline 30 \end{array} \qquad \begin{array}{r} 41 \\ +30 \\ \hline \text{71, 71 cm} \end{array}$$

4.

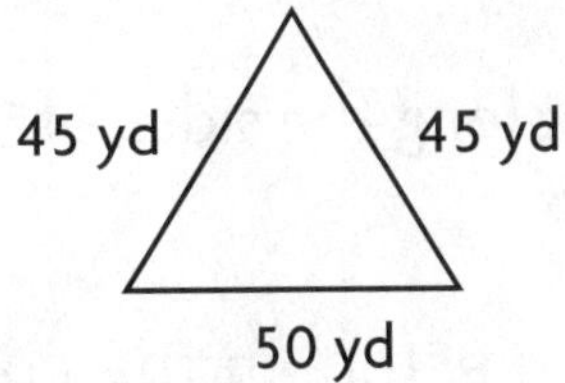

45 + 45 + 50 = __?__ yd

$$\begin{array}{r} 45 \\ +45 \\ \hline 90 \end{array} \qquad \begin{array}{r} 0 \\ +50 \\ \hline 50 \end{array} \qquad \begin{array}{r} 90 \\ +50 \\ \hline \text{140, 140 yd} \end{array}$$

Name ______________________

LESSON 1.3

Problem-Solving Strategy

Make a Model

You will need paper clips and counters. Read the problem and then follow the directions. One paper clip equals one yard.

> Mr. Taft's sixth-grade class is setting up the gym for a dance. The perimeter of the gym is 80 yards. It is in the shape of a rectangle. The gym is 24 yards long. How wide is it?

Fill in what you know about the dance floor.

1. perimeter **80 yd**
2. number of sides **4**
3. length of each of 2 sides **24 yd**

Take 80 paper clips. Build the 2 sides whose lengths you know.

4. How many paper clips are left? **32**

Take the remaining paper clips. Build 2 sides of equal length.

5. How wide is the gym?

   16 yd

> The decoration committee is going to arrange clusters of balloons along the perimeter of the dance floor. There will be 4 clusters of balloons on each side, including 1 at each corner.

Place counters along the sides of your model to represent balloon clusters.

6. Count the clusters in the corners. How many? **4 clusters**
7. Count the clusters not in the corners. How many? **8 clusters**
8. How many balloon clusters will be placed in all?

   12 clusters in all

Name ______________________

Estimating Sums and Differences

Round 63 to the nearest ten.

You can use a number line to help you.

- Mark 63 on the number line.
- Circle the two nearest tens.
- Decide which ten 63 is closer to. Round to that number.
- Since 63 is closer to 60, you round to 60.

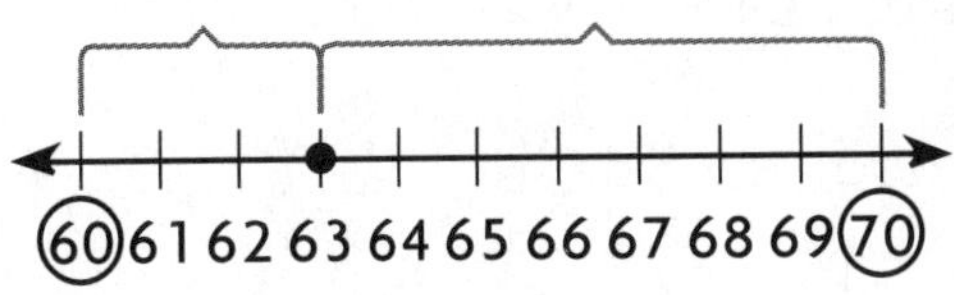

Remember: When a number is halfway between two numbers, round to the greater number.

Round to the nearest ten. Use the number line.

1. 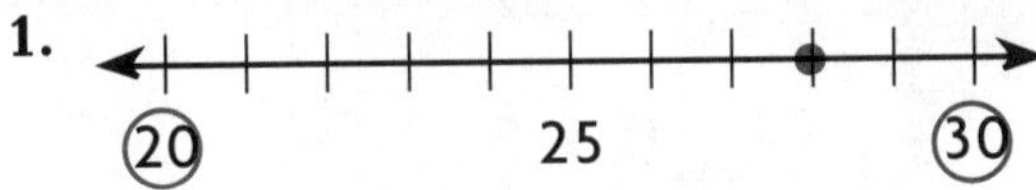

28 → 30

2. 
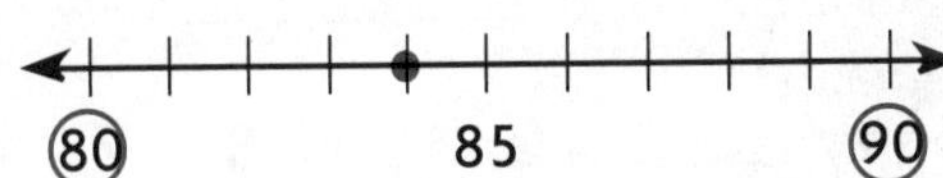

84 → 80

3. 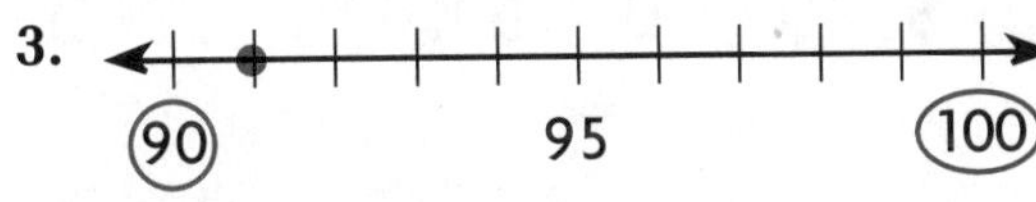

91 → 90

4. 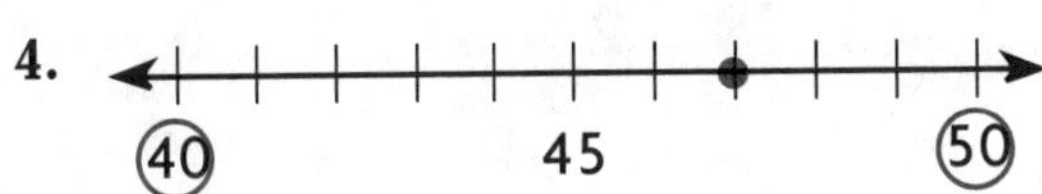

47 → 50

Round to the nearest hundred. Use the number line.

5. 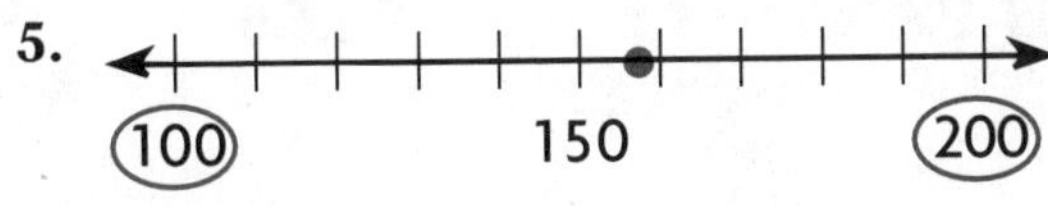

159 → 200

6. 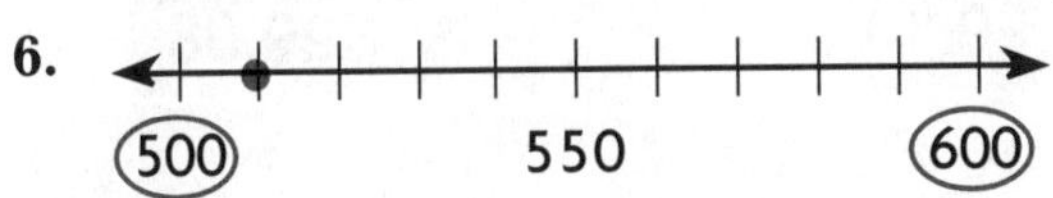

510 → 500

Estimate the sum or difference. Use the rounded numbers from Exercises 1–6.

7. $28 \rightarrow 30$; $+47 \rightarrow +50$; ? → 80

8. $91 \rightarrow 90$; $-84 \rightarrow -80$; ? → 10

9. $47 \rightarrow 50$; $+91 \rightarrow +90$; ? → 140

10. $84 \rightarrow 80$; $-28 \rightarrow -30$; ? → 50

11. $159 \rightarrow 200$; $+510 \rightarrow +500$; ? → 700

12. $510 \rightarrow 500$; $-159 \rightarrow -200$; ? → 300

Name ______________________

LESSON 1.5

Adding and Subtracting with Money

When you write money amounts,

- leave two places for cents after the decimal point.
- place a dollar sign in front of the money amount.

925 → 9.25 → $ 9.25

Write the money amount.

A. 805 → $ 8.05

1. 346 → $ 3.46

2. 100 → $ 1.00

3. 598 → $ 5.98

B. 10 → $ 0.10

4. 65 → $ 0.65

5. 43 → $ 0.43

6. 17 → $ 0.17

C. 1482 → $ 14.82

7. 2337 → $ 23.37

8. 1164 → $ 11.64

9. 2815 → $ 28.15

D. 6 → $ 0.06

10. 3 → $ 0.03

11. 5 → $ 0.05

12. 9 → $ 0.09

Find the sum or difference.

13. $5.41 + 3.72 = **$9.13**

14. $17.66 + 11.42 = **$29.08**

15. $9.83 − 7.91 = **$1.92**

16. $24.15 − 18.60 = **$5.55**

17. $12.87 + 7.09 = **$19.96**

18. $20.50 − 5.29 = **$15.21**

19. $23.75 − 10.99 = **$12.76**

20. $2.29 + 3.05 + 1.81 = **$7.15**

Look at Exercises 13–20. Circle *yes* or *no.*

21. There are two places for cents after each decimal point. 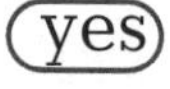 yes no

22. There is a dollar sign in front of each money amount. 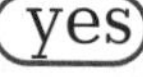 yes no

Name ______________________________

Adding and Subtracting Larger Numbers

What is the length of the fourth side of this figure?

You can use addition *and* subtraction to find the missing length of a figure.

- Add the known sides.
- Subtract the sum from the perimeter.

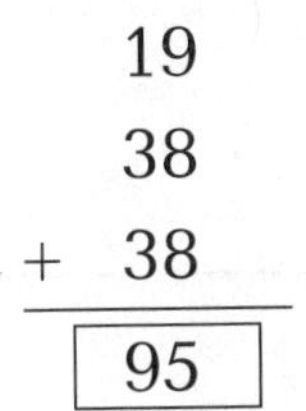

$$\begin{array}{r} 143 \\ -\ \boxed{95} \\ \hline 48 \end{array}$$

? = 48 ft

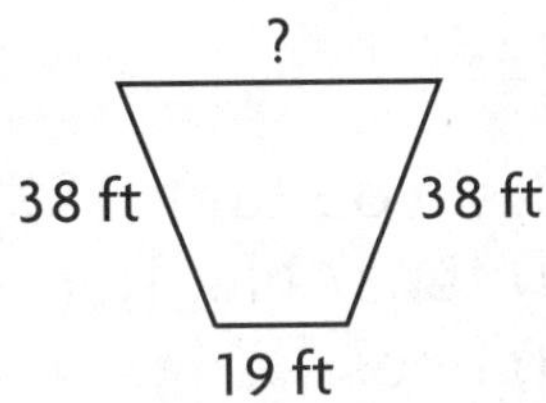

The length of the fourth side is 48 feet.

Perimeter = 143 ft

Use addition *and* subtraction to find the missing length.

1. 124 yd; 102 yd; 102 yd; ?

Perimeter = 404 yd

$$\begin{array}{r} 124 \\ 102 \\ +\ 102 \\ \hline \boxed{328} \end{array} \qquad \begin{array}{r} 404 \\ -\ \boxed{328} \\ \hline 76 \end{array}$$

? = 76 yd

2. ?; 29 ft; 29 ft

Perimeter = 87 ft

$$\begin{array}{r} 29 \\ +\ 29 \\ \hline \boxed{58} \end{array} \qquad \begin{array}{r} 87 \\ -\ \boxed{58} \\ \hline 29 \end{array}$$

? = 29 ft

3. 24 m; 48 m; ?; 48 m; 24 m

Perimeter = 202 m

$$\begin{array}{r} 48 \\ 48 \\ 24 \\ +\ 24 \\ \hline \boxed{144} \end{array} \qquad \begin{array}{r} 202 \\ -\ \boxed{144} \\ \hline \boxed{58} \end{array}$$

? = 58 m

4. ?; 32 ft; 16 ft

Perimeter = 88 ft

? = 40 ft

5. 88 m; 33 m; ?; 88 m

Perimeter = 242 m

? = 33 m

6.

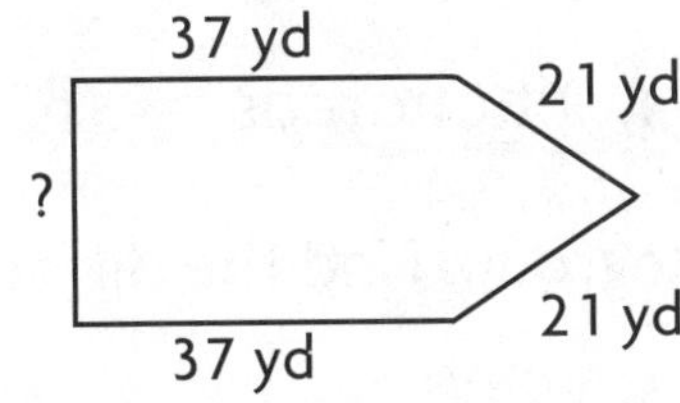

Perimeter = 140 yd

? = 24 yd

Name ________________________________

LESSON 2.2

Subtracting Across Zeros

To subtract across zeros, it is helpful to make a model with base-ten blocks.

Can you take 127 from 5 hundreds? No. Regroup 1 hundred as 10 tens.

$$\begin{array}{r} 500 \\ -127 \\ \hline \end{array}$$

Can you take 127 from 4 hundreds, 10 tens? No. Regroup 1 ten as 10 ones.

$$\begin{array}{rrr} 4 & 10 & \\ \not{5} & \not{0} & 0 \\ -1 & 2 & 7 \\ \hline \end{array}$$

Can you take 127 from 4 hundreds, 9 tens, and 10 ones? Yes.

$$\begin{array}{rrr} & 9 & \\ 4 & \not{10} & 10 \\ \not{5} & \not{0} & \not{0} \\ -1 & 2 & 7 \\ \hline 3 & 7 & 3 \end{array}$$

Hundreds	Tens	Ones
5 hundreds		
4 hundreds	10 tens	
4 hundreds	9 tens	10 ones

Complete Exercises 1–4. Regroup 1 hundred as 10 tens and 1 ten as 10 ones.

1. 3 hundreds = **2** hundreds **9** tens **10** ones
2. 6 hundreds = **5** hundreds **9** tens **10** ones
3. 4 hundreds = **3** hundreds **9** tens **10** ones
4. 9 hundreds = **8** hundreds **9** tens **10** ones

Regroup. Find the difference.

5. $\begin{array}{r} 300 \\ -158 \\ \hline \mathbf{142} \end{array}$ 6. $\begin{array}{r} 600 \\ -149 \\ \hline \mathbf{451} \end{array}$ 7. $\begin{array}{r} 400 \\ -325 \\ \hline \mathbf{75} \end{array}$ 8. $\begin{array}{r} 900 \\ -276 \\ \hline \mathbf{624} \end{array}$

Name ______________________________

LESSON 2.3

More Subtracting Across Zeros

When you add or subtract numbers with many digits, it's important that you write the numbers carefully so that digits with the same place value line up. This also helps you write regrouped digits in the correct places.

	3	9	9	9	10
	4	0,	0	0	0
−	3	2,	9	6	3
		7,	0	3	7

Copy each exercise into the space below. Find the difference. Remember to leave the first row of boxes empty for regrouping.

1. 900 − 831

	8	9	10
	9	0	0
−	8	3	1
		6	9

2. 5,002 − 2,764

	4	9	9	12
	5,	0	0	2
−	2,	7	6	4
	2,	2	3	8

3. 7,001 − 3,208

	6	9	9	11
	7,	0	0	1
−	3,	2	0	8
	3,	7	9	3

4. 6,000 − 1,955

	5	9	9	10
	6,	0	0	0
−	1,	9	5	5
	4,	0	4	5

5. 80,000 − 31,523

	7	9	9	9	10
	8	0,	0	0	0
−	3	1,	5	2	3
	4	8,	4	7	7

6. 10,000 − 4,661

		9	9	9	10
	1	0,	0	0	0
−		4,	6	6	1
		5,	3	3	9

7. 40,000 − 20,118

	3	9	9	9	10
	4	0,	0	0	0
−	2	0,	1	1	8
	1	9,	8	8	2

8. 70,000 − 55,049

	6	9	9	9	10
	7	0,	0	0	0
−	5	5,	0	4	9
	1	4,	9	5	1

9. 60,000 − 33,604

	5	9	9	9	10
	6	0,	0	0	0
−	3	3,	6	0	4
	2	6,	3	9	6

Name ____________________

LESSON 2.3

Problem-Solving Strategy

Work Backward

Sometimes you can work backward to solve a problem.

Bruce collected some sand dollars. He gave 10 to his younger sister. His friend, Susan, gave him 7 sand dollars. He put 4 back in the water. He had 14 sand dollars left. How many did he collect?

Operation	Opposite Operation
Add	Subtract
Subtract	Add
Divide	Multiply
Multiply	Divide

- First, write the steps of the problem in order from left to right. Use numbers and operations to describe the steps.

Number of sand dollars collected.		He gave 10 to sister.		Susan gave him 7.		He put 4 back.		Number left
?	→	− 10	→	+ 7	→	− 4	→	14

- Then, reverse the order and use the opposite operations. Find the answer.

21	←	+ 10	←	− 7	←	+ 4	←	14

So, Bruce collected 21 sand dollars.

Reverse the order and use opposite operations. Find the answer.

1. ? → + 3 → − 8 → − 11 → 27

   43 ← − 3 ← + 8 ← + 11 ← 27

2. ? → − 11 → + 25 → 52

   38 ← + 11 ← − 25 ← 52

3. Sasha and Chuck played a number game. Sasha picked a number and subtracted 3. Next, she added 12. Then, she subtracted 20. Last, she added 11. The result was 30. What number did Sasha begin with?

   30

Name ____________________

Estimating Sums and Differences

An estimate is an answer that is close to the exact answer.

Round 4,782 to the nearest *thousand.*

Follow these rules:

- Circle the digit to be rounded. (4),782
- Underline the digit to its right. (4),7̲82
- If the underlined digit is 5 or more, the circled digit increases by 1. If the underlined digit is less than 5, the circled digit stays the same. 7 > 5, 5,xxx
- Write zeros in all places to the right of the circled digit. 5,000

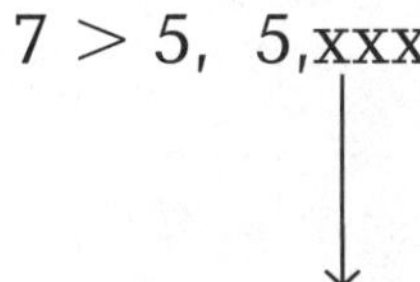

So, 4,782 rounded to the nearest thousand is 5,000.

Circle the digit to be rounded. Underline the digit to its right.
Round 9,385 to the given place value.

1. nearest ten

9,385

9,390

2. nearest hundred

9,385

9,400

3. nearest thousand

9,385

9,000

Round 2,617 to the given place value.

4. nearest ten

2,617

2,620

5. nearest hundred

2,617

2,600

6. nearest thousand

2,617

3,000

Round each number to the greatest place value. Write the estimated sum or difference.

7.

891	→	900
− 623	→	− 600
?	→	300

8.

3,456	→	3,000
+ 7,099	→	+ 7,000
?	→	10,000

Name ____________________

LESSON 2.5

Choosing the Operation

In deciding if you should add or subtract to solve a problem, think about the answer. Should the answer be greater than the numbers in the problem? If yes, then consider adding. If no, then consider subtracting.

Eldrick bought a book for $6.95. He gave the clerk $10.00. What should his change be?

Think: Should the answer be more than or less than $10.00?

less than

Solve:

```
  $10.00
−   6.95
  $3.05
```

Read the problem. Answer the **Think** question. Then solve.

1. Devon bought a card for $1.75 and a magazine for $2.95. How much did he spend?

   Think: Should the answer be more than or less than $2.95?

   more than

   Solve:

   ```
     $1.75
   + 2.95
     $4.70
   ```

2. Alicia paid $8.25 for a calendar with horse pictures on it. Becky paid $11.50 for a calendar with cat pictures on it. How much more did Becky pay than Alicia?

   Think: Should the answer be more than or less than $11.50?

   less than

   Solve:

   ```
     $11.50
   −  8.25
     $3.25
   ```

3. Scott had $15.00 when he went to the store. He has $4.39 left. How much money did he spend at the store?

   Think: Should the answer be more than or less than $15.00?

   less than

   Solve:

   ```
     $15.00
   −   4.39
     $10.61
   ```

Name ________________________________

What Questions Can Multiplication Answer?

When groups *are not* equal, you add to find how many. When groups *are* equal, you can add or multiply to find how many.

Rita bought 3 packages of batteries. There were 2 batteries in each package. How many batteries did Rita buy?

Draw a picture: (||) (||) (||)

Equal groups? yes

Add: 2 + 2 + 2 = 6

Multiply: 3 × 2 = 6

Rita bought 3 packages of batteries. Two packages had 2 batteries each. The third package had 3 batteries. How many batteries did Rita buy?

Draw a picture: (||) (||) (|||)

Equal groups? no

Add: 2 + 2 + 3 = 7

Multiply: cannot

Decide if the groups are equal. Add to find the sum. Multiply if you can. If you cannot multiply, write *cannot.*

1. 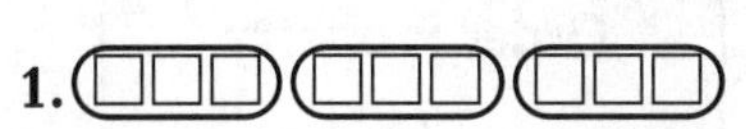

Equal groups? yes

Add: 3 + 3 + 3 = 9

Multiply: 3 × 3 = 9

2. 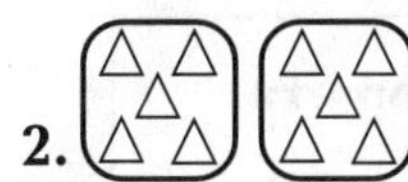

Equal groups? yes

Add: 5 + 5 = 10

Multiply: 2 × 5 = 10

3. 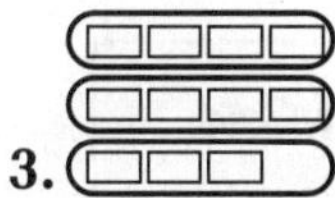

Equal groups? no

Add: 4 + 4 + 3 = 11

Multiply: cannot

Draw a picture. Add to find the sum. Multiply if you can. If you cannot multiply, write *cannot.*

4. Alberto handed out crayons. He gave Pete 5 crayons, Max 7 crayons, and Leslie 8 crayons. How many crayons did Alberto hand out? Draw a picture:

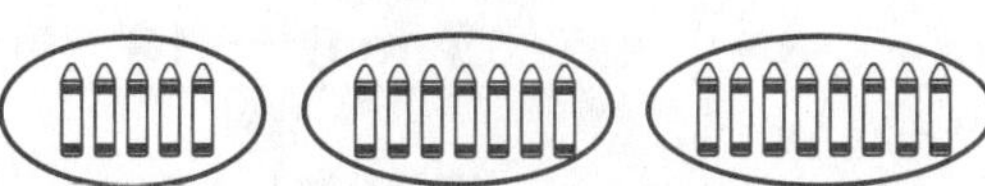

Add: 5 + 7 + 8 = 20

Multiply: cannot

5. Jyoti handed out crayons. She gave 5 crayons to each student in the first row. There are 4 students in the first row. How many crayons did Jyoti hand out? Draw a picture:

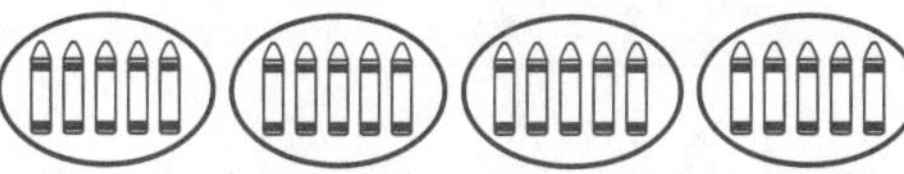

Add: 5 + 5 + 5 + 5 = 20

Multiply: 4 × 5 = 20

Name ______________________________

LESSON 3.2

Multiplication Properties

In a multiplication sentence, the numbers being multiplied are the factors; the answer is the product.

2 × 5 = 10	5 × 2 = 10	9 × 0 = 0
factors product	factors product	factors product

Sort. Write each multiplication sentence in the column under the property it shows.

8 × 0 = 0	1 × 5 = 5	4 × 1 = 4	3 × 6 = 18
9 × 0 = 0	0 × 4 = 0	0 × 6 = 0	2 × 0 = 0
1 × 8 = 8	3 × 0 = 0	3 × 1 = 3	5 × 0 = 0
6 × 1 = 6	4 × 5 = 20	0 × 7 = 0	9 × 1 = 9
7 × 2 = 14	1 × 1 = 1	1 × 7 = 7	

1. **Property of One** When one of the factors is 1, the product is the other factor.	2. **Zero Property** When one of the factors is zero, the product is zero.	3. **Order Property** Two numbers can be multiplied in any order. The product is the same.
1 × 1 = 1	0 × 1 = 0	3 × 4 = 12
1 × 2 = 2	2 × 0 = 0	4 × 3 = 12
3 × 1 = 3	3 × 0 = 0	2 × 7 = 14
4 × 1 = 4	0 × 4 = 0	7 × 2 = 14
1 × 5 = 5	5 × 0 = 0	6 × 3 = 18
6 × 1 = 6	0 × 6 = 0	3 × 6 = 18
1 × 7 = 7	0 × 7 = 0	5 × 4 = 20
1 × 8 = 8	8 × 0 = 0	4 × 5 = 20
9 × 1 = 9	9 × 0 = 0	

Order of sentences in first two columns may vary.

Name ______________________________

More About Multiplication Models

You can use the facts for 5 and a model to help you find the product of factors greater than 5. What is 7 × 8?

5 Facts Chart

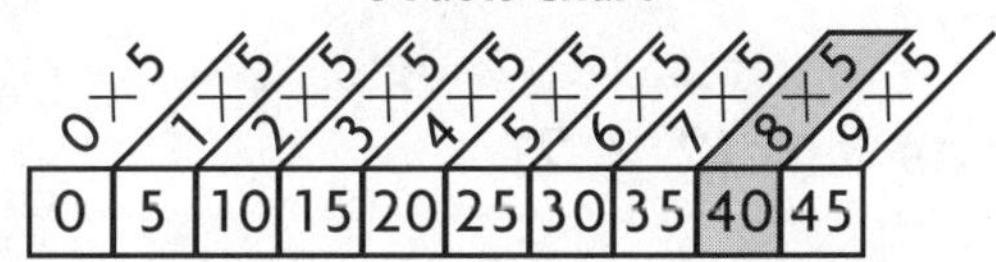

7 × 8 = ?

Step 1 Write an addition expression breaking apart one factor.
5 + 2
(5 × 8) + (2 × 8) = ?

Step 2 Use the 5 Facts Chart and a model to find the products.

Step 3 Find the sum of the products.
40 + 16 = 56

So, 7 × 8 = 56.

Complete each step. Use the 5 Facts Chart to help you draw a model. **Check students' drawings.**

1. 9 × 6 = ?
5 + 4
(5 × 6) + (4 × 6) =
30 + **24**
So, 9 × 6 = **54**

2. 8 × 9 = ?
5 + 3
(5 × 9) + (3 × 9) =
45 + **27**
So, 8 × 9 = **72**

3. 6 × 7 = ?
(5 + 2)
(6 × 5) + (6 × 2) =
30 + **12**
So, 6 × 7 = **42**

4. 8 × 6 = ?
(5 + 3)
(5 × 6) + (3 × 6) =
30 + **18**
So, 8 × 6 = **48**

Name ____________________

LESSON 3.4

Multiplying Three Factors

The **Grouping Property** may help you when you multiply. Remember when the grouping of factors is changed, the product remains the same. Look at the example below which shows the Grouping Property.

$(3 \times 2) \times 2$ gives the same answer as $3 \times (2 \times 2)$.

Choose the way of grouping that is easier for you to multiply. What is $5 \times 2 \times 3$?

- Group the factors two ways. $(5 \times 2) \times 3 =$? or $5 \times (2 \times 3) =$?
- Choose the way that is easier for you.
- Solve.

Think: $5 \times 2 = 10$. It's easy to multiply tens.

$10 \times 3 = 30$ or
$5 \times 6 = 30$

Circle the grouping that is easier for you. Show how you solve it. **Possible answer is shown.**

1. $(2 \times 2) \times 3 =$?, or $2 \times (2 \times 3) =$? (circled)

   $2 \times 6 = 12$

2. $(2 \times 5) \times 3 =$? (circled), or $2 \times (5 \times 3) =$?

   $10 \times 3 = 30$

3. $(4 \times 2) \times 3 =$? (circled), or $4 \times (2 \times 3) =$?

   $8 \times 3 = 24$

4. $(5 \times 3) \times 3 =$?, or $5 \times (3 \times 3) =$? (circled)

   $5 \times 9 = 45$

5. $(2 \times 2) \times 8 =$? (circled), or $2 \times (2 \times 8) =$?

   $4 \times 8 = 32$

6. $(4 \times 3) \times 3 =$?, or $4 \times (3 \times 3) =$? (circled)

   $4 \times 9 = 36$

7. $(5 \times 2) \times 2 =$? (circled), or $5 \times (2 \times 2) =$?

   $10 \times 2 = 20$

8. $(2 \times 6) \times 1 =$?, or $2 \times (6 \times 1) =$? (circled)

   $2 \times 6 = 12$

Name ____________________

Problem-Solving Strategy

Make a Model

In Mr. Paul's classroom there are 3 rows of tables. There are 2 tables in each row. Each table has 4 chairs. How many chairs are in Mr. Paul's classroom?

You can *make a model* to find the number of chairs in the classroom.

Use tiles to make a model.

(3	×	2)	×	4	=	?
number of rows		number of tables		number of chairs at each table		number of chairs in all

Row 1

Row 2

Row 3

$6 \times 4 = 24$

So, there are 24 chairs in Mr. Paul's classroom.

Complete each step. Make a model. Find the product.

Check students' models.

1. Ms. Carter uses a bulletin board to display students' work. The bulletin board has 4 rows of papers on it. There are 5 papers in each row. How many papers are there in all?

   number of rows **4**

   number of papers in each row **5**

   number of bulletin boards **1**

   (**4** × **5**) × **1** = ?

   20 × 1 = 20

   So, there are **20** papers in all.

2. In P.E. class the students are standing in 6 rows. There are 4 students in each row. The P.E. teacher gives 2 hand weights to each student. How many hand weights does she give out?

   number of rows **6**

   number of students in each row **4**

   number of hand weights for each student **2**

   6 × (**4** × **2**) = ?

   6 × 8 = 48

   So, she gives out **48** weights.

Name ______________________________

LESSON 3.5

Choosing the Operation

In deciding if you need to add, multiply, or subtract to solve a problem, think about the answer.

- If the answer should be greater than the numbers in the problem, **add.**
- If the answer should be greater than the numbers in the problem, and the problem includes equal groups, **multiply.**
- If the answer should be less than the greater number in the problem, **subtract.**

Read the problem. Answer the **Think** questions. Then solve by using a number sentence.

1. Rosemary bicycled 6 miles before lunch and 8 miles after lunch. How many miles did she bicycle in all?

   Think: Should the answer be greater than or less than

   8 miles? greater than

   If greater than, does the problem include equal

   groups? no

   Solve: 8 + 6 = 14

   So, Rosemary bicycled 14 mi.

2. Eric had 8 marbles. He gave some to Wendy. Now he has 6 marbles left. How many marbles did he give to Wendy?

   Think: Should the answer be greater than or less than

   8 marbles? less than

   If greater than, does the problem include equal

   groups? ______

   Solve: 8 − 6 = 2

   So, Eric gave Wendy 2 marbles.

Name ______________________________

What Questions Can Division Answer?

You can use division to find how many are in each group or how many equal groups there are. You can use counters to model the division.

Example

Bill is handing out 20 paint-brushes. He places 4 on each table. On how many tables can he place paintbrushes?

total number	20
number in each group	4
number of groups	?

Use counters to find the **number of groups.**

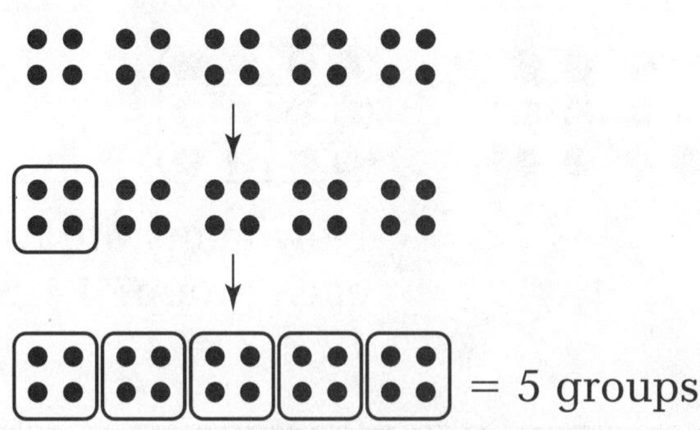

So, he can place paintbrushes on 5 tables.

Write the information from the problem. Write a question mark to show what you need to find. Then use counters to solve.

1. Eva has 24 beads to make 3 bracelets. Each bracelet will have the same number of beads. How many beads will she use for each bracelet?

total number	24
number in each group	?
number of groups	3

   She will use **8 beads**.

2. Pam is gluing photos onto pieces of poster board. She has 27 photos. She is gluing 9 onto each piece of poster board. How many pieces of poster board will she need?

total number	27
number in each group	9
number of groups	?

   She will need **3 pieces**.

Name ____________________

LESSON 4.2

Connecting Multiplication and Division

Multiplication and division are inverse operations. One operation undoes the other. You can use models to show this.

$3 \times 4 = 12$

$12 \div 3 = ?$
Divide 12 into 3 equal groups.

How many are in each group? 4
$12 \div 3 = 4$

$5 \times 2 = 10$

$10 \div 5 = ?$
Divide 10 into 5 equal groups.

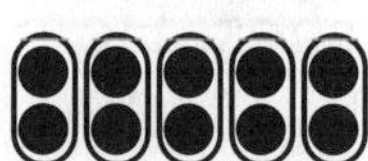

How many are in each group? 2
$10 \div 5 = 2$

Circle equal groups to show the division. Complete the division sentence.

1. $6 \times 3 = 18$

   $18 \div 6 =$ **3**

2. $2 \times 7 = 14$

   $14 \div 2 =$ **7**

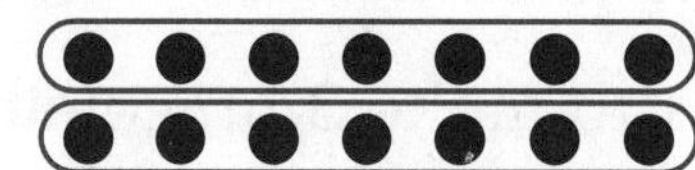

3. $5 \times 4 = 20$

   $20 \div 5 =$ **4**

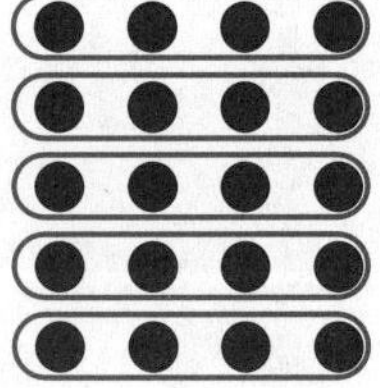

4. $3 \times 9 = 27$

   $27 \div 3 =$ **9**

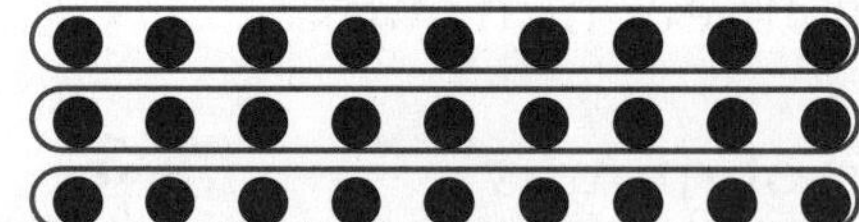

Complete each multiplication and division sentence. You may wish to use a model to help you.

5. $8 \times 6 =$ **48**

   $48 \div 8 =$ **6**

6. $7 \times 4 =$ **28**

   $28 \div 7 =$ **4**

7. $2 \times 9 =$ **18**

   $18 \div 2 =$ **9**

8. $3 \times 3 =$ **9**

   $9 \div 3 =$ **3**

9. $1 \times 8 =$ **8**

   $8 \div 1 =$ **8**

10. $5 \times 7 =$ **35**

    $35 \div 5 =$ **7**

Name ____________________

Dividing with Remainders

Facts of 2	Facts of 3	Facts of 4	Facts of 5	Facts of 6
18 ÷ 2 = 9	27 ÷ 3 = 9	36 ÷ 4 = 9	45 ÷ 5 = 9	54 ÷ 6 = 9
16 ÷ 2 = 8	24 ÷ 3 = 8	32 ÷ 4 = 8	40 ÷ 5 = 8	48 ÷ 6 = 8
14 ÷ 2 = 7	21 ÷ 3 = 7	28 ÷ 4 = 7	35 ÷ 5 = 7	42 ÷ 6 = 7
12 ÷ 2 = 6	18 ÷ 3 = 6	24 ÷ 4 = 6	30 ÷ 5 = 6	36 ÷ 6 = 6
10 ÷ 2 = 5	15 ÷ 3 = 5	20 ÷ 4 = 5	25 ÷ 5 = 5	30 ÷ 6 = 5
8 ÷ 2 = 4	12 ÷ 3 = 4	16 ÷ 4 = 4	20 ÷ 5 = 4	24 ÷ 6 = 4
6 ÷ 2 = 3	9 ÷ 3 = 3	12 ÷ 4 = 3	15 ÷ 5 = 3	18 ÷ 6 = 3
4 ÷ 2 = 2	6 ÷ 3 = 2	8 ÷ 4 = 2	10 ÷ 5 = 2	12 ÷ 6 = 2
2 ÷ 2 = 1	3 ÷ 3 = 1	4 ÷ 4 = 1	5 ÷ 5 = 1	6 ÷ 6 = 1

To find the quotient and remainder, first find the fact whose dividend is closest to yet less than the dividend in the problem. Then subtract the dividends.

What is 23 ÷ 5?

fact: 20 ÷ 5 = 4 subtract: 23 – 20 = 3 So, 23 ÷ 5 = 4 r3.

Use the table and subtract to find the quotient and remainder.

1. What is 27 ÷ 6?

fact: 24 ÷ 6 = 4

subtract: 27 – 24 = 3

So, 27 ÷ 6 = 4 r3.

2. What is 14 ÷ 4?

fact: 12 ÷ 4 = 3

subtract: 14 – 12 = 2

So, 14 ÷ 4 = 3 r2.

3. What is 17 ÷ 3?

15 ÷ 3 = 5

17 – 15 = 2

17 ÷ 3 = 5 r2

4. What is 34 ÷ 5?

30 ÷ 5 = 6

34 – 30 = 4

34 ÷ 5 = 6 r4

5. What is 15 ÷ 2?

14 ÷ 2 = 7

15 – 14 = 1

15 ÷ 2 = 7 r1

Name ____________________

LESSON 4.3

Problem-Solving Strategy

Make a Table

You can make a table to show all the possible ways something can be done. You can use the information in the table to solve problems.

Charise has decorated 36 barrettes to sell at a craft fair. She needs to decide whether to put 4, 5, 6, or 8 barrettes in each package. She made a table to help find out how she might package the barrettes.

Complete the table.

	Total Number of Barrettes	Number in Each Package	Division Sentence	Number of Packages	Number of Barrettes Left Over
	36	4	36 ÷ 4 = 9	9	0
1.	36	5	36 ÷ 5 = 7 r1	7	1
2.	36	6	36 ÷ 6 = 6	6	0
3.	36	8	36 ÷ 8 = 4 r4	4	4

4. If Charise does not want to have any barrettes left over, how many ways can she package the barrettes? What are the ways?

   2 ways; 9 packages of 4 or 6 packages of 6

5. If Charise would like to have 1 barrette left over to keep for herself, how many barrettes should she put in each package?

   5 barrettes

6. How many more barrettes would Charise need to make to put 8 barrettes in each package and have none left over?

   4 more

7. Charise decides she wants to make the fewest packages she can and have no leftover barrettes. How many barrettes should she put in each package?

   6 barrettes

Name ______________________________________

Division on a Multiplication Table

You can use a multiplication table and two index cards to find a quotient. To find the quotient 12 ÷ 4 = __?__, follow these steps.

- Write a related multiplication fact with a missing factor.

  ? × 4 = 12

  So, 12 ÷ 4 = 3.

- Find the given factor in the top row of the chart. Place an index card alongside the column it is in.

×	0	1	2	3	4
0	0	0	0	0	0
1	0	1	2	3	4
2	0	2	4	6	8
3	0	3	6	9	12
4	0	4	8	12	16
5	0	5	10	15	20
6	0	6	12	18	24

- Find the product 12 in that column. Place an index card below. The first number in that row is the answer.

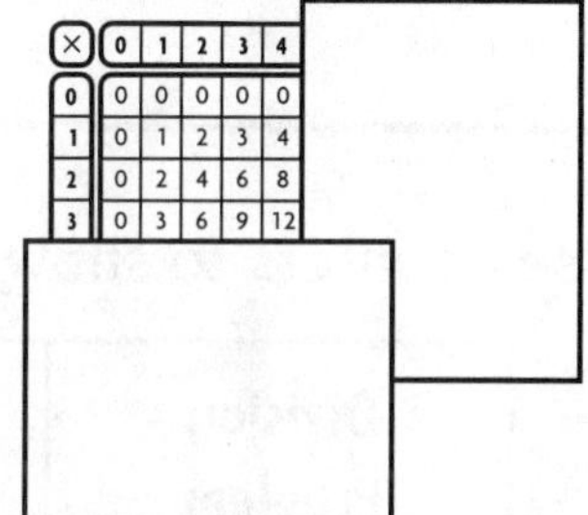

Write the related multiplication sentence. Use the multiplication table to solve.

×	0	1	2	3	4	5	6	7	8	9
0	0	0	0	0	0	0	0	0	0	0
1	0	1	2	3	4	5	6	7	8	9
2	0	2	4	6	8	10	12	14	16	18
3	0	3	6	9	12	15	18	21	24	27
4	0	4	8	12	16	20	24	28	32	36
5	0	5	10	15	20	25	30	35	40	45
6	0	6	12	18	24	30	36	42	48	54
7	0	7	14	21	28	35	42	49	56	63
8	0	8	16	24	32	40	48	56	64	72
9	0	9	18	27	36	45	54	63	72	81

1. 27 ÷ 3 = __?__

  ? × 3 = 27

  27 ÷ 3 = **9**

2. 42 ÷ 7 = __?__

  ? × 7 = 42

  42 ÷ 7 = **6**

3. 24 ÷ 6 = __?__

  ? × 6 = 24

  24 ÷ 6 = **4**

Use the multiplication table to find the quotient.

4. 36 ÷ 4 = **9** **5.** 48 ÷ 8 = **6** **6.** 20 ÷ 5 = **4**

7. 28 ÷ 7 = **4** **8.** 56 ÷ 7 = **8** **9.** 45 ÷ 9 = **5**

10. 36 ÷ 6 = **6** **11.** 63 ÷ 9 = **7** **12.** 81 ÷ 9 = **9**

Name ______________________

LESSON 4.5

Recording and Practicing Division

You can record division problems in two ways: $6 \div 1 = ?$ or $1\overline{)6}$ with ? as the quotient.

How many counters? 6

How many groups? 1

How many in each group? 6

So, $6 \div 1 = 6$ or $1\overline{)6}$ with quotient 6.

Use counters to show the division. Complete each table.

	Division Problem	Number of Counters	Number of Groups	Number in Each Group	Division Sentence
1.	$4 \div 4 = ?$	4	4	1	$4 \div 4 =$ 1
2.	$6 \div 6 = ?$	6	6	1	$6 \div 6 =$ 1
3.	$7 \div 7 = ?$	7	7	1	$7 \div 7 =$ 1
4.	$8 \div 8 = ?$	8	8	1	$8 \div 8 =$ 1

5. When a number is divided by itself, what is the quotient? 1

	Division Problem	Number of Counters	Number of Groups	Number in Each Group	Division Sentence
6.	$\overset{?}{1\overline{)3}}$	3	1	3	$\overset{3}{1\overline{)3}}$
7.	$\overset{?}{1\overline{)5}}$	5	1	5	$\overset{5}{1\overline{)5}}$
8.	$\overset{?}{1\overline{)8}}$	8	1	8	$\overset{8}{1\overline{)8}}$
9.	$\overset{?}{1\overline{)9}}$	9	1	9	$\overset{9}{1\overline{)9}}$

10. When a number is divided by 1, what is the quotient? that number

Name ______________________________

LESSON 4.6

Choosing the Operation

Decide if you should add, multiply, subtract, or divide to get the answer.

- If the answer should be greater than the numbers in the problem, **add.**
- If the answer should be greater than the numbers in the problem, and the problem includes equal groups, **multiply.**
- If the answer should be less than the greater number in the problem, **subtract.**
- If the answer should be less than the greater number in the problem, and the problem includes equal groups, **divide.**

Read the problem. Answer the **Think** questions. Choose the operation, and then solve.

1. Lewis bought 6 packs of astronomy cards. Each pack had 7 cards in it. How many cards did Lewis buy?

Think: Should the answer be greater than or less than 7 cards? greater than

Does the problem include equal groups?

yes

Solve: $6 \times 7 = 42$

Lewis bought 42 cards.

2. Hugh, Carol, and Lee collected 24 pinecones. They divided them so that they each got the same number. How many pinecones did they each get?

Think: Should the answer be greater than or less than 24 pinecones? less than

Does the problem include equal groups?

yes

Solve: $24 \div 3 = 8$

They each got 8 pinecones.

Name ______________________________

How Numbers Are Used

Different kinds of numbers give us different information.

Cardinal numbers tell how many.	**Ordinal numbers** tell position or order.	**Nominal numbers** name things.
• **1** inch	• **4th** grade	• radio station: **103.1** FM
• **two** hours	• **ninth** birthday	• telephone number: **555-4027**
• **6** students	• **first** anniversary	• identification number: **007**
• **10** acorns	• **18th** floor	
• **fifty-two** miles		

In the diagram below, color each cardinal number blue, each ordinal number yellow, and each nominal number green.

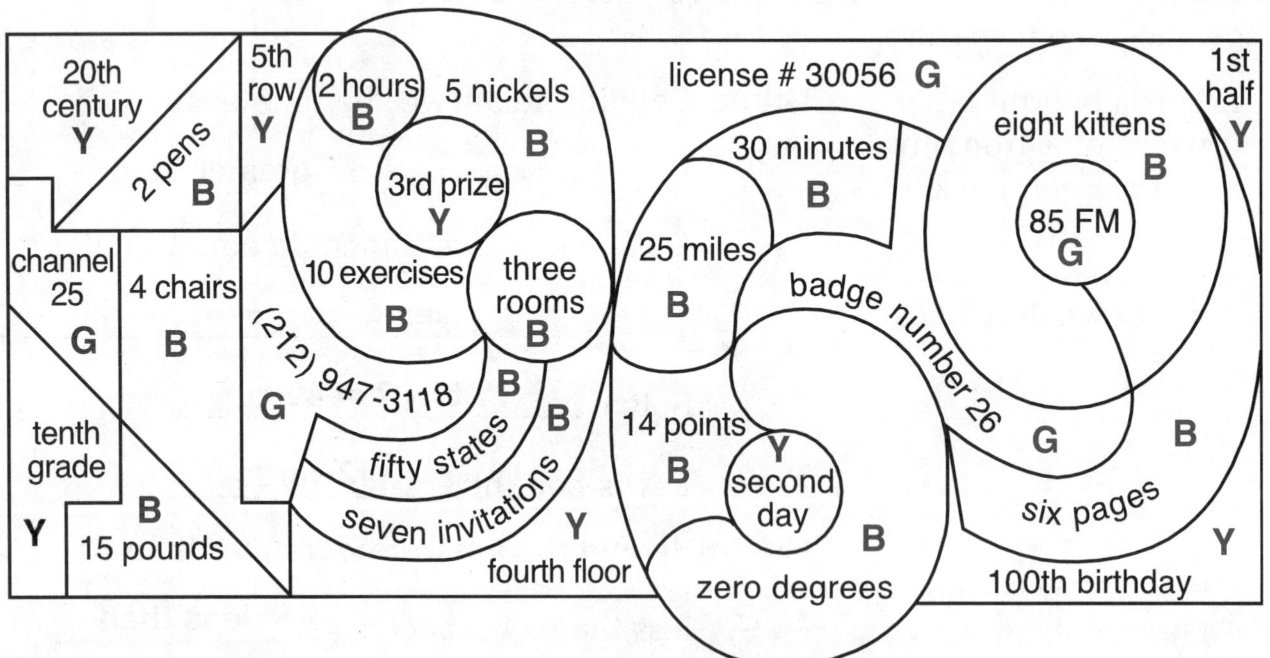

In what year did the first astronaut walk on the moon? Look at your picture for the answer. 1969

Name ___

More About How Numbers Are Used

The two numbers used to locate points on a grid are called an **ordered pair.**

What is the ordered pair for the location of the Ferris wheel?

(_?_, _?_)

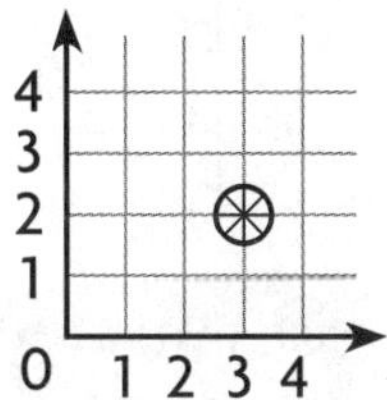

⊛ = Ferris wheel

- Locate the Ferris wheel. Follow the gridline **down** to a number. That is the first number in the ordered pair.

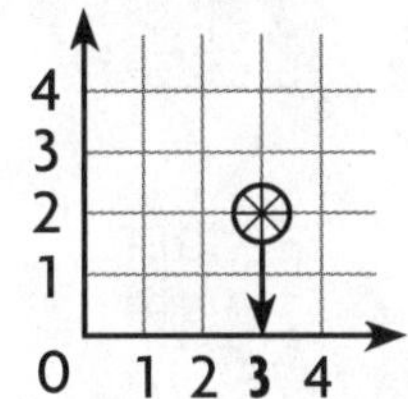

(3, _?_)

- Follow the gridline **across** to a number. That is the second number in the ordered pair.

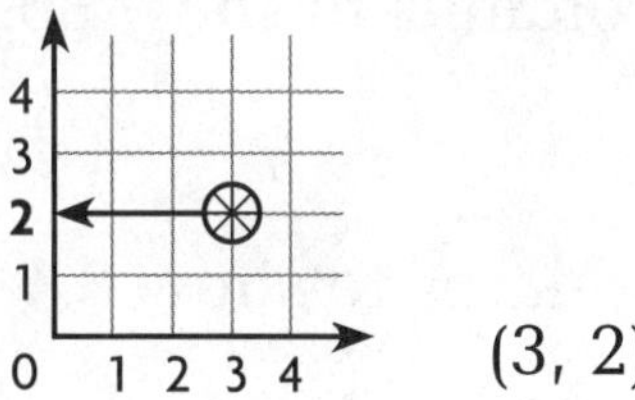

(3, 2)

So, (3, 2) is the ordered pair for the location of the Ferris wheel.

Complete the ordered pair for the location of the Ferris wheel.

1.

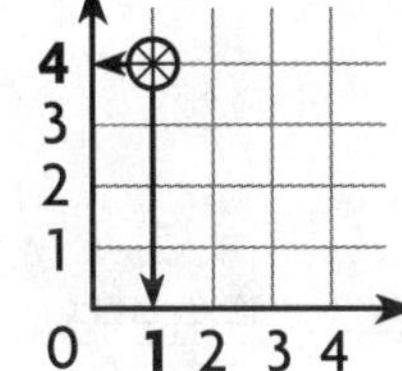

(1, __4__)

2.

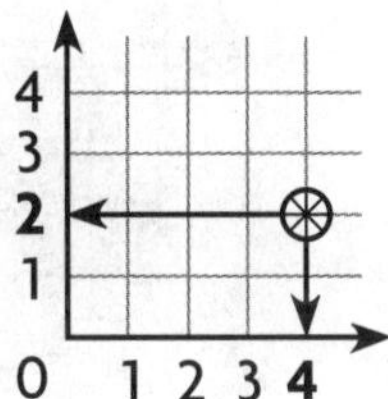

(__4__, 2)

3.

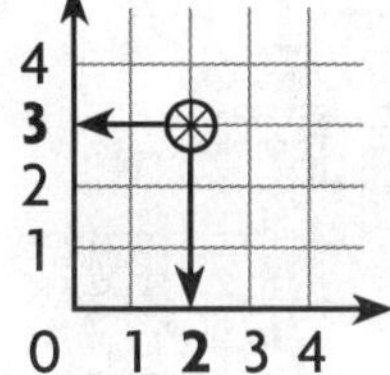

(__2__, __3__)

Write the ordered pair for each ride.

4. merry-go-round ____(4, 3)____

5. bumper cars ____(2, 1)____

6. roller coaster ____(2, 5)____

7. tea cups ____(6, 5)____

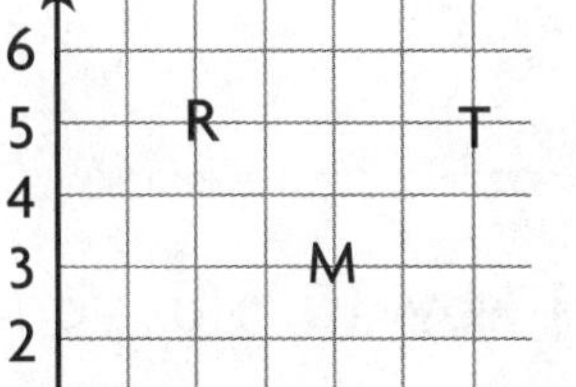

Legend

R = roller coaster

B = bumper cars

M = merry-go-round

T = tea cups

Name ______________________

LESSON 5.2

Understanding 1,000

You can represent numbers from 1 to 1,000 using 10-by-10 grids.

Remember: ▢ = 1 = 10 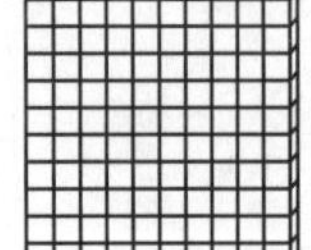 = 100

How many 10-by-10 grids do you need to show 438?

Show 438. **Think:** 438 = 400 + 30 + 8

- Shade grids to show 400.

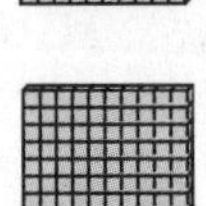 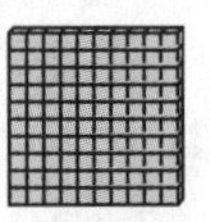 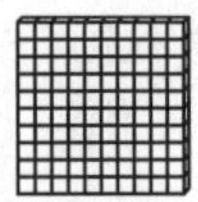

- Shade columns to show 30.

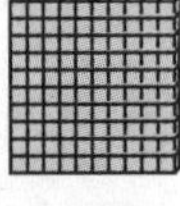 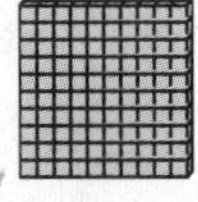  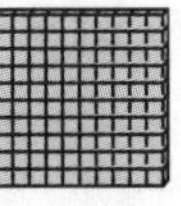 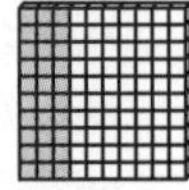

- Shade squares to show 8.

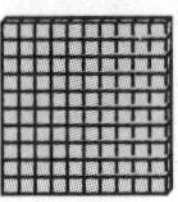 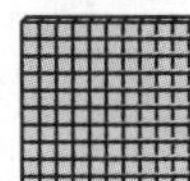 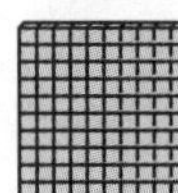 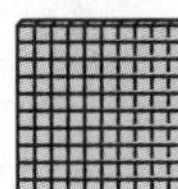 

So, you need 5 grids to show 438.

Write the number. Tell how many 10-by-10 grids are needed. (Grids that are overlapped are completely shaded.)

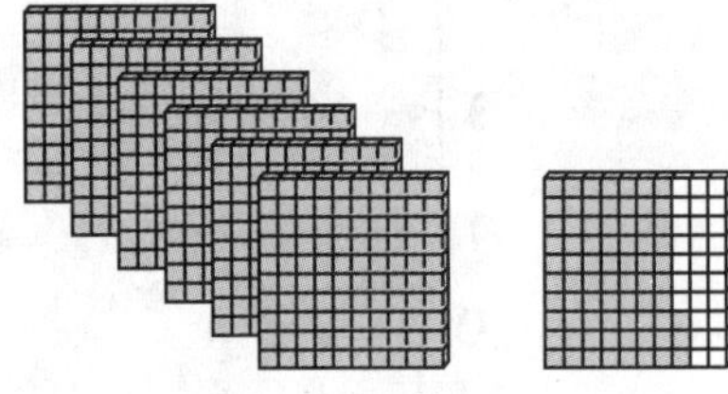

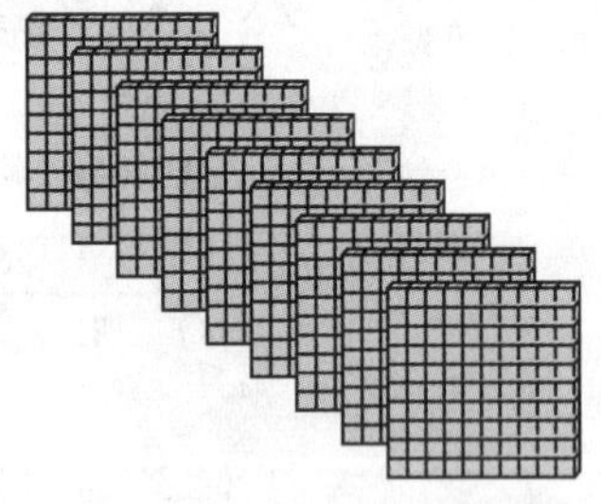 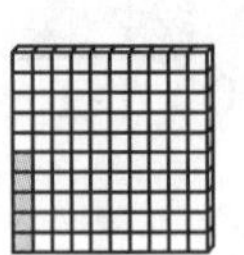

1. number 673

   number of grids 7

2. number 905

   number of grids 10

Write the number of 10-by-10 grids you would need to show the number.

3. 80 — 1 grid
4. 800 — 8 grids
5. 808 — 9 grids
6. 961 — 10 grids

Name ______________________________

More About 1,000

In order to read and write numbers to thousands, it is important to know the value of each digit in a number.

You can write the numbers 6,507 and 2,089 in words.

• Write each digit in its place in a place-value chart.	Thousands: 6, Hundreds: 5, Tens: 0, Ones: 7	Thousands: 2, Hundreds: 0, Tens: 8, Ones: 9
• Write the number in expanded form.	6,000 + 500 + 7	2,000 + 80 + 9
• Write the number in words. Remember to write a comma after *thousand*.	six thousand, five hundred seven	two thousand, eighty-nine

---

Write each number in expanded form. Then write each number in words.

1. 8,021 = **8,000** + **20** + **1**

   eight thousand, twenty-one

2. 4,617 = **4,000** + **600** + **10** + **7**

   four thousand, six hundred seventeen

3. 1,938 = **1,000** + **900** + **30** + **8**

   one thousand, nine hundred thirty-eight

4. 6,472 = **6,000** + **400** + **70** + **2**

   six thousand, four hundred seventy-two

5. 9,706 = **9,000** + **700** + **6**

   nine thousand, seven hundred six

Name ______________________________

LESSON 5.3

Problem-Solving Strategy

Act It Out

Laverne, Jody, Pearl, Leroy, and Cal are each performing in a variety show. Laverne will perform after Leroy. Pearl will perform third. Jody will perform before Pearl. Cal will perform before Jody. In what order will they perform?

Act it out to solve. Write each person's name on a slip of paper. When you know a person's position, place the slip of paper with his or her name in that position.

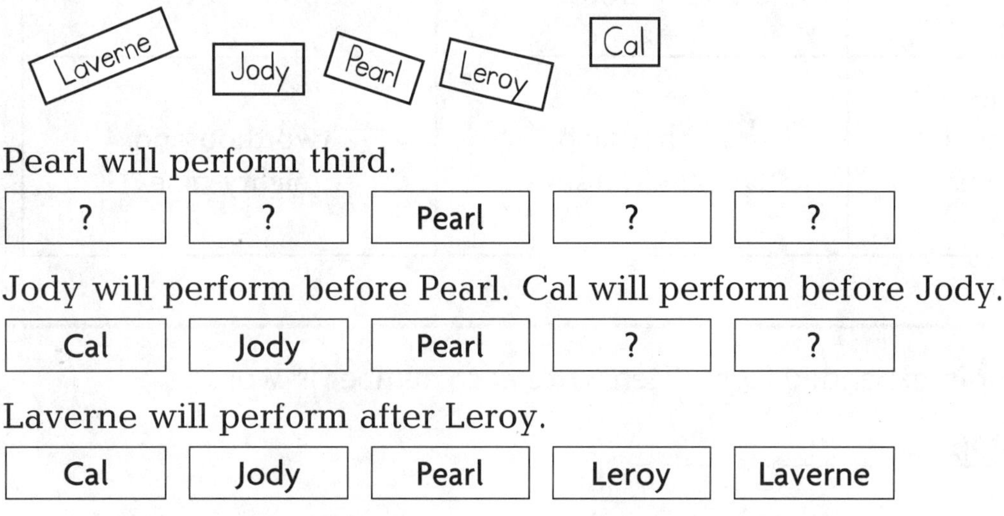

Pearl will perform third.

?	?	Pearl	?	?

Jody will perform before Pearl. Cal will perform before Jody.

Cal	Jody	Pearl	?	?

Laverne will perform after Leroy.

Cal	Jody	Pearl	Leroy	Laverne

So, the order is: Cal, Jody, Pearl, Leroy, Laverne.

Act it out to solve.

1. Vicky has 5 classes before lunch each day. She has math before science. She has Spanish before math. She has English right before lunch. She has history after science. In what order does Vicky have classes in the morning?

   Spanish, math, science, history, English

2. There are 6 children in the Ingraham family. Denise is older than Steven, but younger than Eliot. Kayla is younger than all her brothers and sisters except for Kirk. Bernie is younger than Steven. Write the names of the Ingraham children in order from oldest to youngest.

   Eliot, Denise, Steven, Bernie, Kayla, Kirk

Name ______________________________

Benchmark Numbers

You can use a benchmark number to help you estimate the number of flowers in gardens *A* and *B*.

- Determine how much space 5 flowers fill. This is your benchmark number.

- Estimate how many times larger is the space filled by flowers in each garden.

A

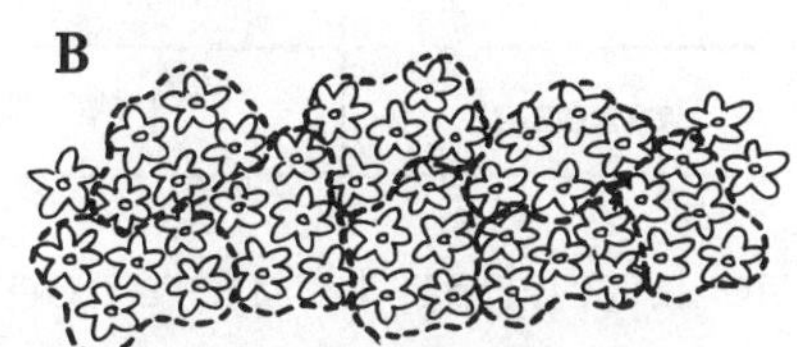

B

The space is about 4 times larger.
$4 \times 5 = 20$

The space is about 8 times larger.
$8 \times 5 = 40$

So, there are about 20 flowers in garden *A* and about 40 flowers in garden *B*.

Complete each problem to estimate the number of seeds shown.
Answers may vary. One possible answer is shown.

1.

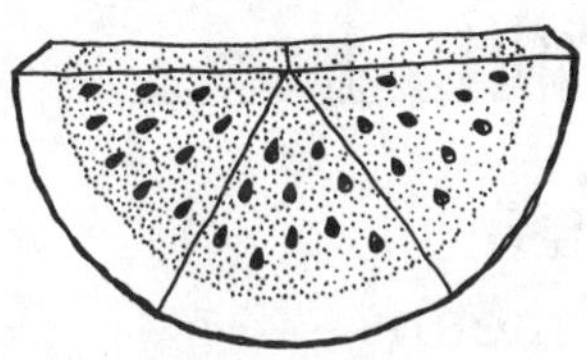

benchmark number **10**

3 × **10** = **30**

about **30** seeds.

2.

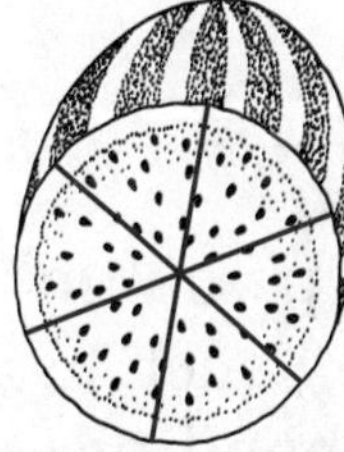

benchmark number **10**

6 × **10** = **60**

about **60** seeds.

Estimate the number of buds on the branches.

3.

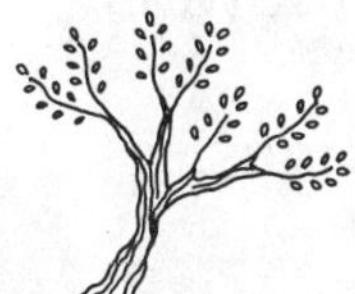

Accept answers of about 50 buds.

4.

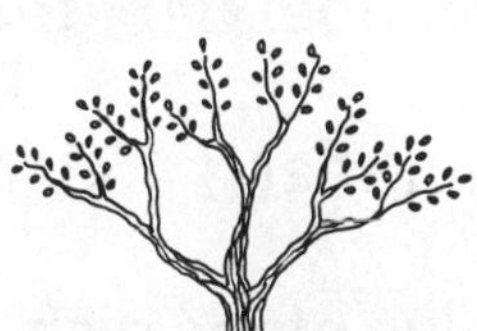

Accept answers of about 80 buds.

Name ______________________________

Numeration Systems

A numeration system is a way of naming numbers and counting numbers. The people of ancient Egypt used these symbols to name numbers.

Egyptian	Lotus Flower	Scroll	∩ Heel Bone	| Stick
Standard	1,000	100	10	1

The Egyptian numeration system did not use place value, so the position and order of the symbols did not matter.

For example, the number ∩∩∩|| and the number ∩∩||∩ both equal the standard number 32.

Write the value of each group of Egyptian symbols.

1. ∩∩||| 23

2. |||∩∩ 23

3. (scroll)(scroll)(scroll)(scroll)|| 402

4. (lotus flower)(scroll)∩∩|||| 1,124

Use a table to help you understand the value of each digit in a large number, such as 3,457. The number 3,457 means 3 thousands + 4 hundreds + 5 tens + 7 ones.

Thousands		Hundreds	Tens	Ones
3	,	4	5	7

Complete.

5. In the number 346, the value of the 3 is 3 hundreds.

6. In the number 3,459, the value of the 9 is 9 ones.

Write the value of the digit 5 in each of the following numbers.

7. 502 500

8. 365 5

9. 5,403 5,000

10. 3,450 50

Name ______________________________

LESSON 6.2

Reading and Writing Numbers

You can show numbers using base-ten blocks, or write numbers in different ways.

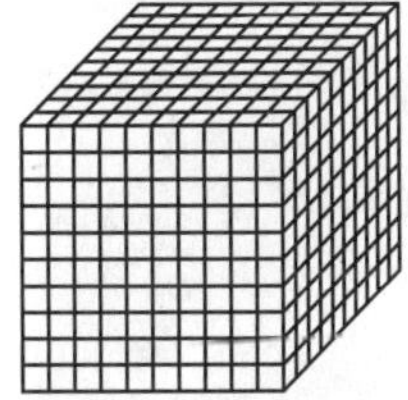
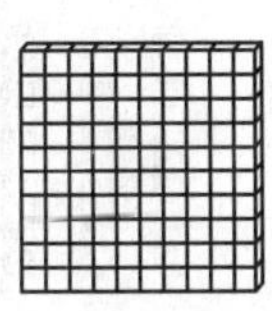
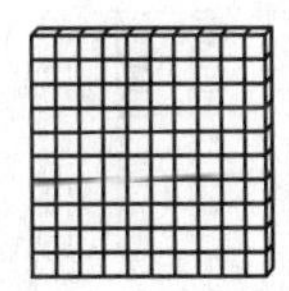
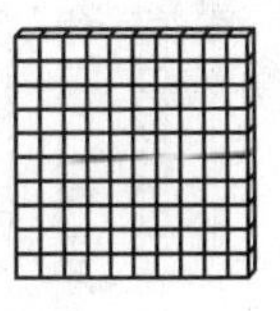
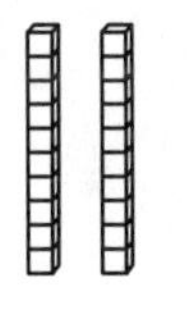

1 thousand 3 hundreds 2 tens 5 ones

Expanded form 1,000 + 300 + 20 + 5

Standard form 1,325

Written form one thousand, three hundred twenty-five

Use the model to help you write the number in different ways.

1. 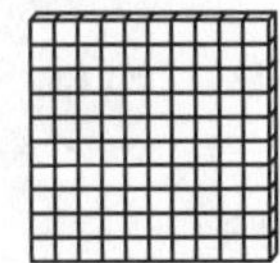
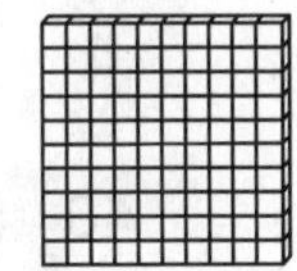
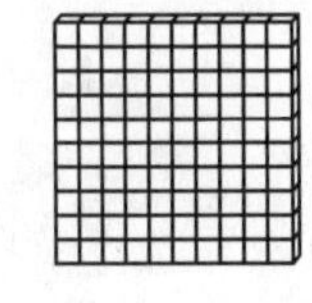
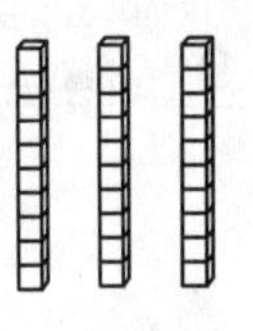

Expanded form 400 + 30 + 8

Standard form 438

Written form four hundred thirty-eight

2. 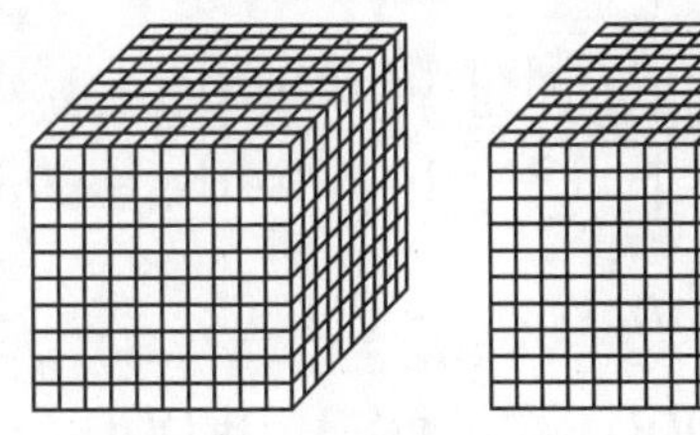
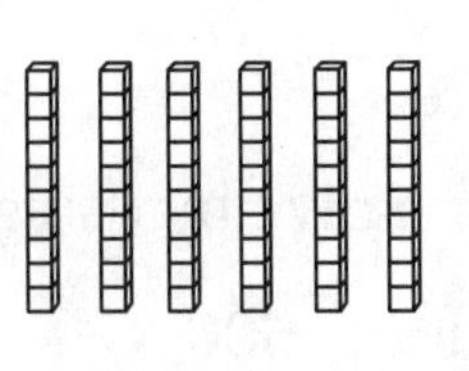

Expanded form 2,000 + 60 + 4

Standard form 2,064

Written form two thousand, sixty-four

Write each number in expanded form.

3. 541 500 + 40 + 1

4. 5,333 5,000 + 300 + 30 + 3

Name ____________________

LESSON 6.3

Mental Math and Place Value

You can use place value to find different names for the same number.

Find two other names for 140.

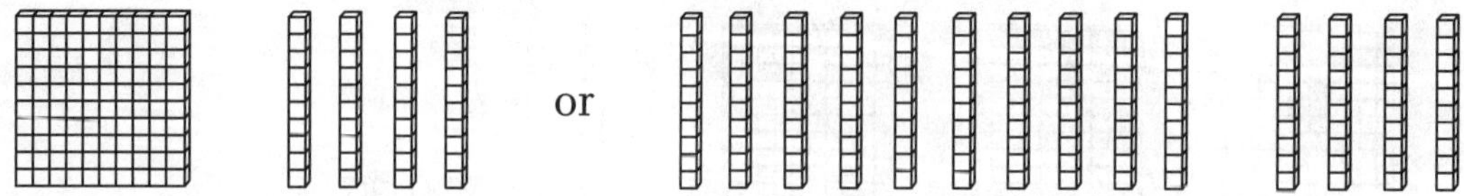

140 = **1** hundred + 4 tens   or   140 = **14** tens

Write each number two ways.

1. 160 = **1** hundred **6** tens
   160 = **16** tens

2. 180 = **1** hundred **8** tens
   180 = **18** tens

3. 290 = 2 hundreds **9** tens
   290 = **29** tens

4. 400 = **4** hundreds
   400 = **40** tens

Use another name for each number. Solve Exercises 5–6 by using mental math.

5. 80 + 90 = ___?___

80 →	8	tens
+90 →	+ 9	tens
	17	tens = 170

6. 700 + 600 = ___?___

700 →	7	hundreds
+600 →	+ 6	hundreds
	13	hundreds = 1,300

Use another name for each number. Solve by using mental math.

7. $30 + 60 = $ **90**

8. $400 + 800 = $ **1,200**

9. $600 + 600 = $ **1,200**

10. $9{,}000 + 2{,}000 = $ **11,000**

11. $5{,}000 + 8{,}000 = $ **13,000**

7. 3 tens + 6 tens = 9 tens **Answers for numbers 8–11 should follow the same style.**

Name ______________________

Place Value to 100,000

In our place-value system, each place is ten times greater than the place to its right.

Hundred Thousands	Ten Thousands	Thousands	Hundreds	Tens	Ones
100,000	10,000	1,000	100	10	1
10 ten thousands	10 thousands	10 hundreds	10 tens	10 ones	1 one

- What is the next largest place after hundreds?
  Think: It is ten times greater than hundreds, so it must be thousands.
- What is the next largest place after thousands?
  Think: It is ten times greater than thousands, so it must be ten thousands.

Complete the place-value chart for the number 111,111.

	Hundred Thousands	Ten Thousands	Thousands	Hundreds	Tens	Ones
1.	100,000	10,000	1,000	100	10	1

Complete the following equations.

2. 10 ten thousands = 100,000
3. 10 hundreds = 1,000
4. 10 thousands = 10,000
5. 10 ones = 10
6. 10 tens = 100
7. 10 hundreds = 1,000
8. ten thousands = 10,000
9. 100 thousands = 100,000

Name ______________________________

LESSON 6.5

Using Large Numbers

Large numbers are separated into periods by commas. A period is a group of three numbers in a large number. The commas make the number easier to read.

Millions				Thousands				Ones		
Hundreds	Tens	Ones		Hundreds	Tens	Ones		Hundreds	Tens	Ones
	4	3	,	3	0	4	,	2	3	5

Read: 43 million, 304 thousand, 235

Write: 43,304,235

Complete to show how to read each number.

1. 45,234,500 ___45___ million, ___234___ thousand, ___500___
2. 8,345,688 ___8___ million, ___345___ thousand, ___688___
3. 240,555,678 240 ___million___, 555 ___thousand___, ___678___
4. 34,111,800 34 ___million___, 111 ___thousand___, ___800___

Write each number in standard form.

5. 43 thousand, 345 ___43,345___
6. 24 million, 234 thousand, 100 ___24,234,100___
7. 5 million, 8 thousand, 34 ___5,008,034___

Write the digit that is in the ten thousands place.

8. 456,100 ___5___
9. 1,290,000 ___9___
10. 123,450,356 ___5___

Write the digit that is in the hundred thousands place.

11. 256,100 ___2___
12. 9,876,000 ___8___
13. 300,123,666 ___1___

Write the digit that is in the millions place.

14. 5,000,123 ___5___
15. 346,123,000 ___6___
16. 301,000,000 ___1___

Name ______________________

Use a Table

A table is helpful in organizing data. Find 1940 under the *Year* column on the table. Look across the table to find the number of farm workers for that year.

Persons in Farm Occupations in the United States, 1940–1990	
Year	Number of Farm Workers
1940	8,995,000
1950	6,858,000
1960	4,132,000
1970	2,881,000
1980	2,818,000
1990	2,864,000

For Problems 1–4, use the table.

1. Were there more farm workers in the United States in 1940 or in 1990? **1940** How many more? **6,131,000**

2. Between which two decades did the number of farm workers increase? **1980** and **1990**

   What was the amount of this increase? **46,000**

3. Do you think that there were more or fewer farm workers in 1930 than in 1940? **more**

   Why? **The number of farm workers is generally decreasing.**

4. After reading the table, Carlos predicted that the number of farm workers in the year 2000 would be about 2,800,000.

   Is this a reasonable prediction? **yes**

   Why or why not? **The number of farm workers has been close to 2,800,000 in 1970, 1980, and 1990.**

Name ________________________________

LESSON 7.1

Comparing on a Number Line

Locate the numbers 20 and 50 on the number line by drawing a dot for each number.

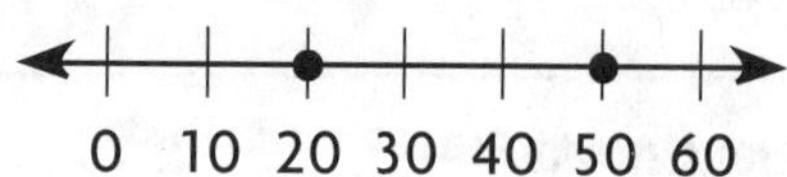

You can compare 20 and 50 by writing 20 < 50 or 50 > 20.

Locate each pair of numbers on the number line by drawing dots. Then show two ways to compare the numbers.

1. 70, 90

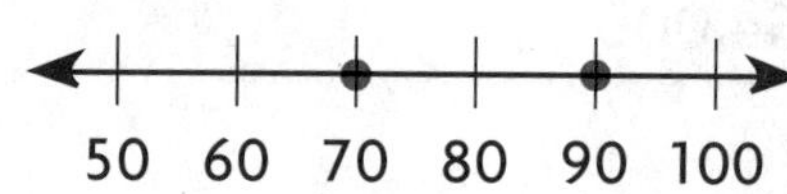

70 < 90 ____ 90 > 70 ____

2. 600, 300

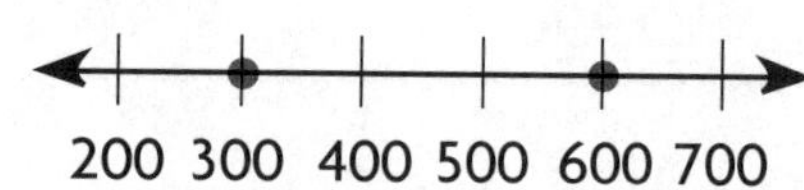

300 < 600 ____ 600 > 300 ____

3. 4,000, 2,000

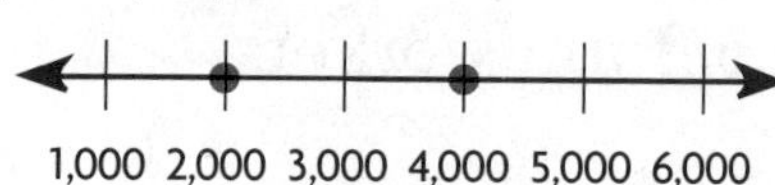

2,000 < 4,000 ____ 4,000 > 2,000 ____

4. 25, 50

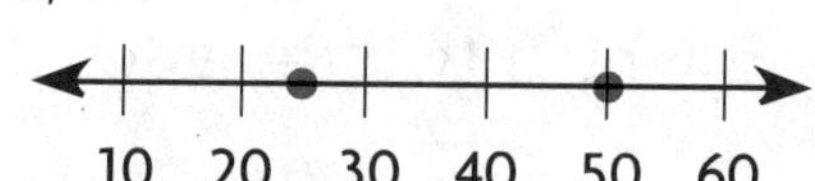

25 < 50 ____ 50 > 25 ____

5. 34, 37

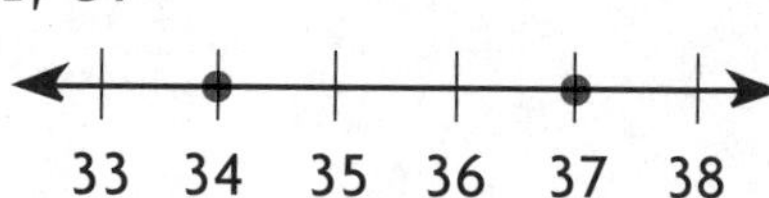

34 < 37 ____ 37 > 34 ____

6. 240, 220

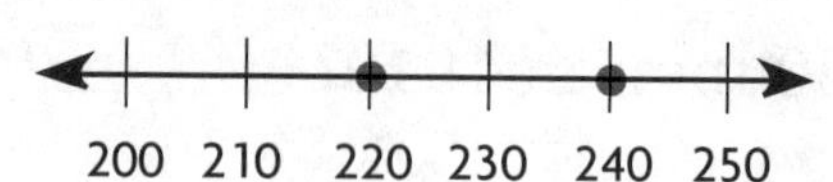

220 < 240 ____ 240 > 220 ____

7. 3,500, 3,200

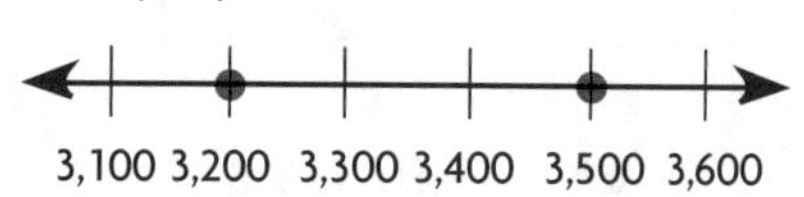

3,200 < 3,500 ____ 3,500 > 3,200 ____

8. 750, 500

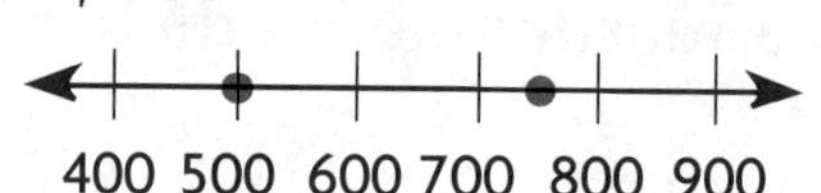

500 < 750 ____ 750 > 500 ____

Name ______________________________

Comparing Numbers

Which number is greater, 3,491 or 3,487?

Step 1 Write one number under the other.

Step 2 Compare the digits, beginning with the greatest place-value position.

Step 3 Circle the first digits that are different.

Think: Since 8 tens < 9 tens,
3,487 < 3,491.

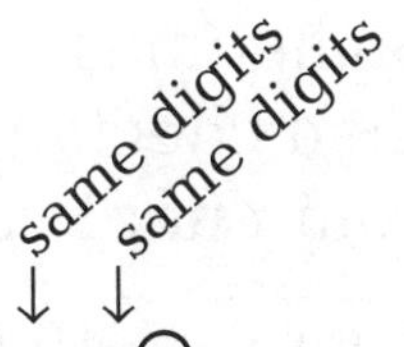

Circle the greater number in each pair of numbers. Then write < or >.

1. 627 (637)	**2.** (129) 127	**3.** 599 (600)
627 (<) 637	129 (>) 127	599 (<) 600
4. 1,345 (1,444)	**5.** 2,400 (3,200)	**6.** (5,225) 4,988
1,345 (<) 1,444	2,400 (<) 3,200	5,225 (>) 4,988
7. (3,010) 3,009	**8.** (1,456) 998	**9.** 890 (894)
3,010 (>) 3,009	1,456 (>) 998	890 (<) 894
10. 7,203 (7,320)	**11.** 4,408 (4,423)	**12.** (763) 739
7,203 (<) 7,320	4,408 (<) 4,423	763 (>) 739
13. 410 (503)	**14.** (1,062) 1,028	**15.** 5,301 (5,401)
410 (<) 503	1,062 (>) 1,028	5,301 (<) 5,401

Name ______________________________

LESSON 7.2

Problem-Solving Strategy

Guess and Check

Sometimes you can find an answer by guessing first and then checking your answer. Then guess, check, and guess again until you get the answer.

There are 16 lions and tigers at the zoo. There are 4 more lions than tigers. How many lions and how many tigers are there?

	Tigers	Lions	Total	Think
Guess 1	4	8	12	not enough animals
Guess 2	7	11	18	too many animals
Guess 3	6	10	16	just right

Use the table to help you guess and check. You may need to guess more than 3 times.

1. There are 24 plants in Tamara's garden. There are twice as many tomato plants as pepper plants. How many tomato and pepper plants are there?

   8 pepper plants, 16 tomato plants

	Pepper Plants	Tomato Plants	Total
Guess 1			
Guess 2			
Guess 3			

2. Leon and Jim have 45 model racing cars. Leon has 5 more cars than Jim. How many cars does each boy have?

   Jim has 20 cars. Leon has 25 cars.

	Jim's Cars	Leon's Cars	Total
Guess 1			
Guess 2			
Guess 3			

Name ______________________

Ordering Numbers

Cut pieces of adding-machine paper the same length as each of the following:

- Your foot
- Your hand (from wrist to longest fingertip)
- Your arm (from shoulder to longest fingertip)
- Distance around your ankle

Label each piece of paper with the name of the body part.

Compare and order the pieces of paper from the least to the greatest. List the measurements in order from the least to the greatest. **Answers will vary.**

1. ______________ 2. ______________

3. ______________ 4. ______________

The Jones family records the height of each family member on New Year's Day. One year, they made this table.

Name	Mrs. Jones	Bill	Lisa	Mr. Jones	Tom	Jeff
Height (in inches)	66 in.	54 in.	39 in.	71 in.	35 in.	48 in.

5. List the heights in order from the greatest to the least.

   71 in., 66 in., 54 in., 48 in., 39 in., 35 in.

6. How much taller was Bill than Tom? **19 in.**

7. How much more did Tom need to grow to be 3 feet tall? **1 in.**

8. What was the difference in height between the tallest and the shortest family members? **36 in.**

9. One year later, Lisa was 2 inches taller and Jeff was 1 inch taller. What was the difference in their heights then? **8 in.**

Name ______________________________

LESSON 7.4

More About Ordering Numbers

The table shows the heights of three mountains in Vermont.

Mountain	Height in Feet
Ellen	4,135 ft
Camel's Hump	4,083 ft
Abraham	4,052 ft

List the heights of the mountains in order from the tallest to the shortest.

Step 1
Compare the digits in each place-value position, beginning with the thousands.

4,135
↓
4,083
↓
4,052

Step 2
Compare the digits in the next place-value position.

4,135
↓
4,083
↓
4,052

Think: 1 hundred > 0 hundreds, so 4,135 is the greatest number.

Step 3
Compare 4,083 and 4,052.

4,083
↓
4,052

Think: 8 tens > 5 tens, so 4,083 > 4,052.

So, the heights of the mountains in order from the tallest to the shortest are 4,135; 4,083; 4,052.

Circle the greatest number and underline the least number.

1. (738)
689 (underlined)
691

2. 3,670 (underlined)
3,760
(3,761)

3. (9,456)
8,390
8,389 (underlined)

4. (7,803)
7,549
7,201 (underlined)

5. 4,023
(4,068)
4,005 (underlined)

6. 6,260 (underlined)
(6,275)
6,270

Name ______________________________

Sorting and Comparing

The librarian recorded the number of books that each student read during the summer reading program at the library.

Alana	Lily	Mark	Sean	Marco	Molly	Jean	Tom
22 books	7 books	38 books	12 books	9 books	18 books	31 books	6 books

The Venn diagram below shows one way to compare the numbers of books read.

Number of Books Read

Fewer Than 10 Books: 7, 6, 9

More Than 10 Books: 22, 38, 12, 31, 18

1. How many students read fewer than 10 books? 3 students

2. How many students read more than 10 books? 5 students

For Exercises 3–4, use the table above. Insert the numbers where they belong in the Venn diagrams.

3. **Number of Books Read**

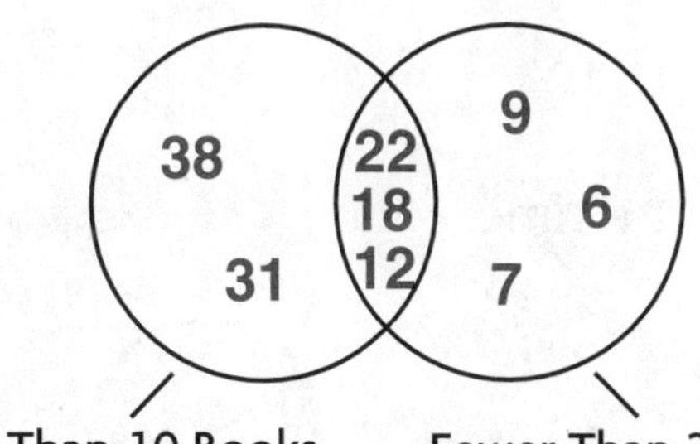

4. **Number of Books Read**

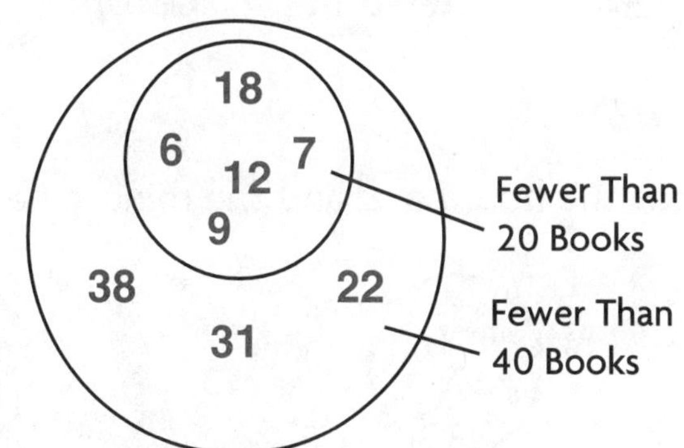

Solve.

5. How many students read more than 10 books but fewer than 25 books? 3 students

6. How many students read fewer than 20 books?

5 students

Name ____________________

Telling Time

Analog Clock

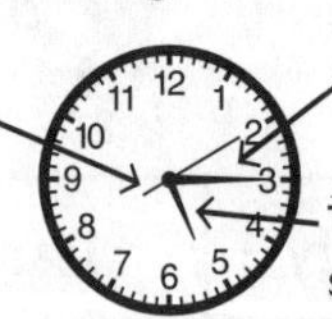

Read: 15 minutes and 10 seconds past 5, or 5:15 and 10 seconds

Digital Clock

Read: 35 minutes past 2

Complete to write the time shown on the clock. Include the seconds.

1. 

25 minutes and 20 seconds past 1

2. 

50 minutes and 45 seconds past 6

3. 

42 minutes and 5 seconds past 4

4. 

8 minutes and 30 seconds past 10

Draw the hour hand and the minute hand to show the time.

5. 
3:25

6. 
8:05

7. 
9:57

Write *second, minute, hour,* or *day*.

8. It took Joe 45 minutes to wash and vacuum his car.

9. A snap of your fingers takes about 1 second.

10. Anne's piano lesson was 1 hour long.

Name ______________________________

A.M. and P.M.

A.M. means "the time between midnight and noon."

At 8:15 A.M. Kathy goes to school.

P.M. means "the time between noon and midnight."

At 8:15 P.M. Kathy goes to bed.

Write A.M. or P.M.

1. George eats supper at 5:30 **P.M.**.
2. Carlos goes to school from 8:40 **A.M.** to 2:40 **P.M.**.
3. Maria has a violin lesson from 3:45 **P.M.** to 4:30 **P.M.**.

Write the time, using A.M. or P.M.

4. the time the library closes

9:00 P.M.

5. the time Kitty sees the sunset

7:25 P.M.

6. the time Lu takes a spelling test at school

10:15 A.M.

7. the time Frank gets home from school

2:55 P.M.

8. the time Jack has soccer practice

4:15 P.M.

9. the time Dad cooks breakfast

6:45 A.M.

Name ______________________________

LESSON 8.3

Elapsed Time on a Clock

Skip counting can help you tell the elapsed time in minutes.
How many minutes is it

from 10:40 to 11:15?

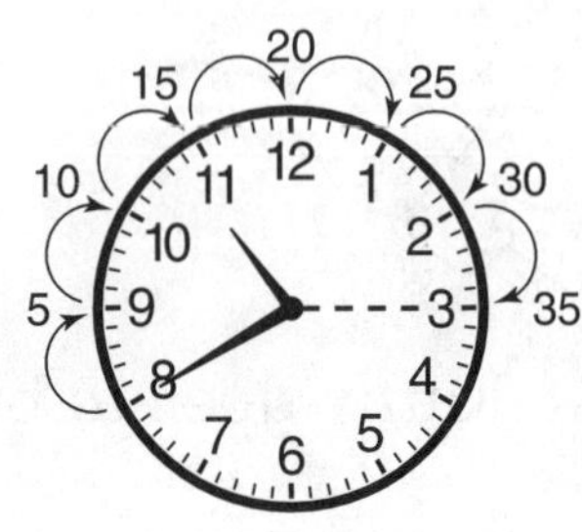

from 6:15 to 7:00?

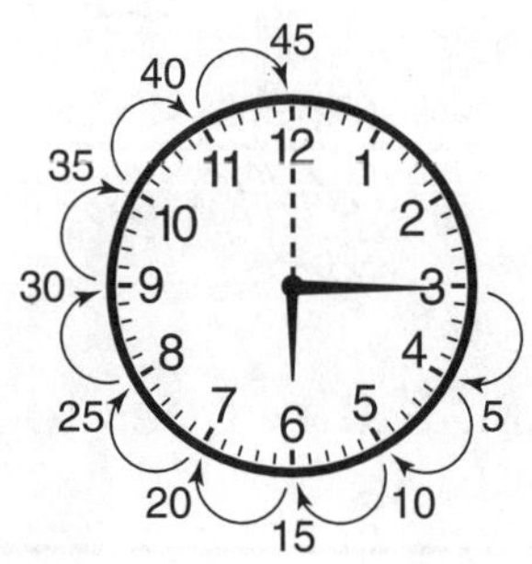

It is 35 minutes from 10:40 to 11:15.

It is 45 minutes from 6:15 to 7:00.

Tell how many minutes will elapse.

1. from 12:30 to 12:45

15 min

2. from 4:25 to 5:20

55 min

3. from 10:15 to 11:05

50 min

4. from 11:15 to 11:40

25 min

5. from 7:05 to 8:00

55 min

6. from 6:35 to 6:45

10 min

Name ______________________

LESSON 8.4

Using a Schedule

WORKSHOP SCHEDULE			
Class	**Days**	**Start Time**	**End Time**
Computer	Tue or Thu	3:30 P.M.	4:15 P.M.
Cooking	Wed	3:00 P.M.	4:00 P.M.
Leather Crafts	Mon or Thu	3:00 P.M.	4:15 P.M.
Drama	Mon	4:00 P.M.	4:45 P.M.
Pottery	Wed	3:15 P.M.	4:15 P.M.

Rosie wants to take a class after school. She is free on Mondays. She needs to take a bus home at 4:30 P.M. What class can she take?

Step 1 Find the column marked *Days*. The leather crafts and drama classes meet on Mondays.

Step 2 Find the column marked *End Time*. Rosie can take the leather crafts class because it ends before 4:30 P.M.

For Problems 1–7, use the workshop schedule.

1. Which classes last 1 hour? **cooking, pottery**

2. Which class is the longest? **leather crafts**

3. Which classes could you take on Thursday?

   computer, leather crafts

4. Could you take both cooking and pottery? **no**

5. John is taking cooking and computer. How long does he attend classes? **1 hr 45 min**

6. It takes Bill 20 minutes to walk home after pottery class. When does he arrive at home? **4:35 P.M.**

7. One day Jill's father picked her up after leather crafts class. He arrived at 4:55 P.M. How long did Jill have to wait for her father? **40 min**

Name ______________________

LESSON 8.4

Problem-Solving Strategy

Make a Table

Saturday Movie Times at the Library

Time Travel will be shown at 9:30 A.M., 11:00 A.M., and 2:15 P.M. Movie length is 45 minutes.

Spaceship Adventure will be shown at 10:30 A.M., 1:15 P.M., and 4:00 P.M. Movie length is 2 hours.

Chen Wong would like to see both movies. He needs to be home by 2:30 P.M.

1. Complete the table, and then circle the times Chen Wong can see each movie. Remember to include A.M. and P.M.

Name of Movie	Starting Times	Ending Times
Time Travel	(9:30 A.M.)	10:15 A.M.
	11:00 A.M.	11:45 A.M.
	2:15 P.M.	3:00 P.M.
Spaceship Adventure	(10:30 A.M.)	12:30 P.M.
	1:15 A.M.	3:15 P.M.
	4:00 P.M.	6:00 P.M.

For Problems 2 and 3, use the table.

2. Bill walked home from the library after watching the 10:30 A.M. showing of *Spaceship Adventure*. It took him 30 minutes to walk home. When did Bill arrive home?

   1:00 P.M.

3. Maria was at the library from 2:00 P.M. to 4:00 P.M. During that time, she watched *Time Travel* and worked on a report. How much time did she spend working on her report? 1 hr 15 min

Name ______________________________

Elapsed Time on a Calendar

You can use a calendar to find elapsed time. How many days will the Harvest Festival last?

- Look at the October calendar. Find 9.
- Count each day of the festival.

So, the festival will last 10 days.

Wilson Farm began advertising the Harvest Festival 5 weeks before the first day of the festival. On what date did Wilson Farm begin advertising?

- Begin at October 9. It is a Thursday.
- Use the calendars to count back 5 weeks, or 5 Thursdays.

So, Wilson Farm began advertising September 4.

For Problems 1–4, use the calendars.

1. Write the date of the first Sunday during the festival. Oct 12
2. Tom bought a pumpkin on the last day of the festival. How many days will it be from then until October 31?

   13 days
3. Julian went to the festival on October 13 after school. What day of the week was that? Mon
4. Wilson Farm began harvesting pumpkins 2 weeks before the beginning of the festival. On what date did the farm begin harvesting pumpkins? Sep 25

September

Sun	Mon	Tue	Wed	Thu	Fri	Sat
	1	2	3	4	5	6
7	8	9	10	11	12	13
14	15	16	17	18	19	20
21	22	23	24	25	26	27
28	29	30				

October

Sun	Mon	Tue	Wed	Thu	Fri	Sat
			1	2	3	4
5	6	7	8	9	10	11
12	13	14	15	16	17	18
19	20	21	22	23	24	25
26	27	28	29	30	31	

Name ______________________________

LESSON 9.1

Organizing Data in Tables

Mr. Cohen used tally marks to keep a record of the number of gerbils that he sold each week in his pet shop.

Week	Gerbils Sold
1	𝍸 𝍸 /
2	𝍸 ///
3	𝍸 ///
4	////

1 =

𝍸 =

In the frequency table, the *frequency* is the number of gerbils Mr. Cohen sold in a week. The *cumulative frequency* is the number of gerbils sold so far.

Use the tally table to complete the frequency table below.

NUMBER OF GERBILS SOLD			
Week	**Gerbils Sold**	**Cumulative Frequency**	
1	**11**	**11**	← sold in Week 1
2	**8**	**19**	← sold in Weeks 1 and 2
3	**8**	**27**	← sold in Weeks 1, 2, and 3
4	**4**	**31**	← sold in Weeks 1, 2, 3, and 4

For Exercises 1–3, use the frequency table to answer the questions.

1. During which week did Mr. Cohen sell the most gerbils?

   Week 1

2. During which two weeks did Mr. Cohen sell the same number of gerbils? **Weeks 2 and 3**

3. How many gerbils did Mr. Cohen sell in Weeks 3 and 4?

   12 gerbils

Name ______________________

Organizing Data

Joanna recorded the results of an experiment. She tossed a coin and a number cube (labeled 1–6) at the same time.

Joanna's Experiment						
	1	2	3	4	5	6
Heads	/	//	///	/	/	///
Tails	/	~~////~~		//	////	//

To see how many times Joanna tossed both heads and 6,

- look across the row labeled *Heads.*
- look down the column labeled 6.

There are three tally marks in this box.

So, Joanna tossed heads and 6 three times.

For Exercises 1–4, use the table above.

1. How many times did Joanna toss both

   heads and 5? **1** tails and 5? **4** tails and 3? **0**

2. How many possible outcomes are there for Joanna's experiment? **12 possible outcomes**

3. How many times did Joanna toss the coin and the number cube? **25 times**

4. Which outcome happened most often? **tails and 2**

5. Toss a penny and a nickel 20 times. Record the results in the table by using tally marks. **Check students' tables. Answers will vary.**

Penny

Nickel		Heads	Tails
	Heads		
	Tails		

Name ______________________________

Problem-Solving Strategy

Make an Organized List

Kyle wants to make a beanbag game to help his younger brother learn about shapes and colors. He will use the colors red, yellow, and blue. He will use circles, squares, and triangles. Kyle made an organized list and drew a picture of each beanbag he can make.

	Red	Yellow	Blue
Circle	R	Y	B
Square	R	Y	B
Triangle	R	Y	B

1. How did Kyle organize his list of possible beanbags?

   He put the shapes in rows and the colors in columns.

2. How many different beanbags can Kyle make using 3 colors and 3 shapes? 9 different beanbags

3. a. Make an organized list by drawing beanbags for a game that uses these colors and shapes: red, blue, yellow, and green; square and circle.

	Red	Blue	Yellow	Green
Square	R	B	Y	G
Circle	R	B	Y	G

   b. How many different beanbags can you make?

   8 different beanbags

Name ________________________________

LESSON 9.3

Understanding Surveys

In a **survey,** you ask several people the same question and record their answers. Mr. Baker made a survey of favorite kinds of ice cream for the class party.

This table shows the results of Mr. Baker's first survey.

For Exercises 1–4, use the First Survey table.

FIRST SURVEY What is your favorite ice cream flavor?	
Flavor	**Votes**
Chocolate	卌 /
Peppermint	卌
Strawberry	卌
Peach	/
Mint Chip	/
Chocolate Chip	///
Vanilla	卌

1. How many students chose vanilla?

   5 students

2. How many flavors of ice cream did the students choose? 7 flavors

3. Which flavor got the most votes?

   chocolate

4. How many students chose either mint chip or chocolate chip as their favorite flavor? 4 students

Mr. Baker wanted to buy just two flavors of ice cream. He made a second survey to decide which two flavors to buy.

SECOND SURVEY Which of these ice cream flavors do you like best—chocolate, strawberry, peppermint, or vanilla?	
Flavor	**Votes**
Chocolate	卌 ///
Strawberry	卌 /
Peppermint	卌 //
Vanilla	卌

5. Why do you think Mr. Baker picked these flavors for his second survey?

   They got the most votes in the first survey.

6. Did more students choose chocolate in the first survey or in the second survey?

   second survey

7. Explain why you think some flavors got more votes in the second survey.

   Possible answer: There were fewer choices in the second survey.

Name ____________________

Comparing Graphs

This vertical bar graph shows how many items were sold one Saturday at the Toy Corner.

How many puzzles were sold?

Follow these steps to help you read the graph.

Step 1 Read across the bottom to find *Puzzles.* Put your finger on the word *Puzzles.*

Step 2 Follow the bar to the top. Place a piece of paper at the top of the bar. Look to the left scale and read.

Step 3 The bar stops halfway between 8 and 10. The number is 9.

There were 9 puzzles sold.

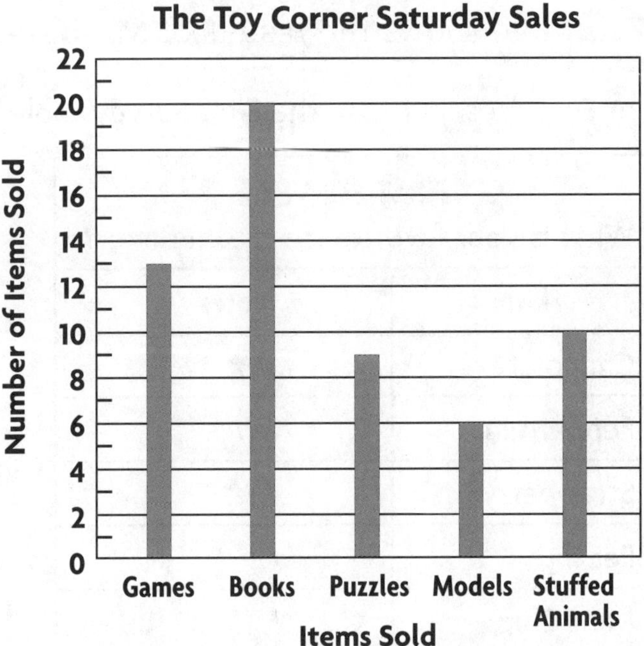

For Exercises 1–10, use the graph.
How many of each item were sold?

1. Games 13
2. Books 20
3. Puzzles 9
4. Models 6
5. Stuffed Animals 10

Which item sold more?

6. Stuffed Animals or Puzzles? Stuffed Animals
7. Games or Models? Games
8. Games or Stuffed Animals? Games

9. How many items were sold in all? 58 items

10. Which two items added together sold almost the same number as books? games and models, or puzzles and stuffed animals

Name ______________________________

More About Comparing Graphs

Tom and his friends earned money cutting lawns during the summer.

Lawns Cut This Summer

Name	Number of Lawns
Tom	6
Jack	10
Barb	8
Kim	4
Lia	3

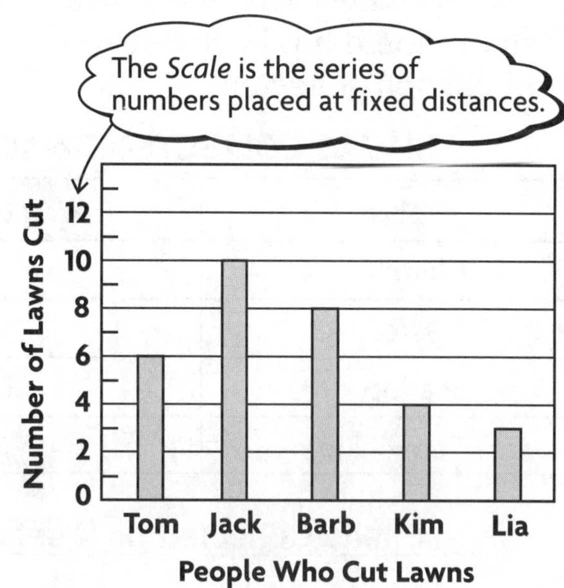

To make a bar graph with different intervals, think about how the graph will look when you are finished. Make two graphs. Follow these steps.

Step 1 Choose an interval for the graph. Label each item and the numbers on the scale.

Step 2 Use a pencil and ruler to measure and fill in the tops, sides, and bottoms of the bars.

Step 3 Check to see if you have correctly filled in the graphs. Color the bars with markers.

Use the table above to complete the bar graphs. Draw a bar to show the number of lawns cut by each person. Use a pencil, ruler, and markers.

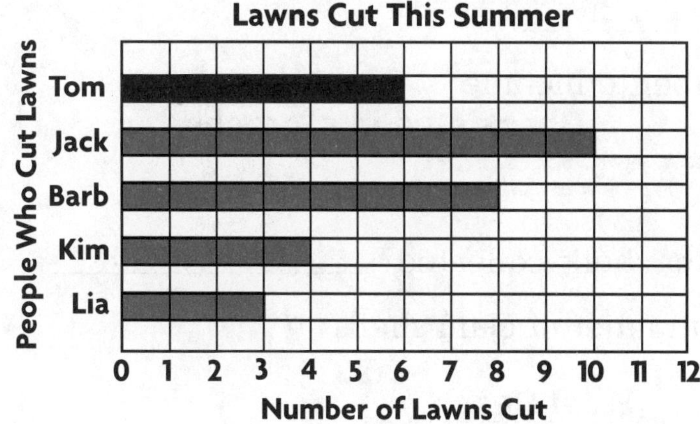

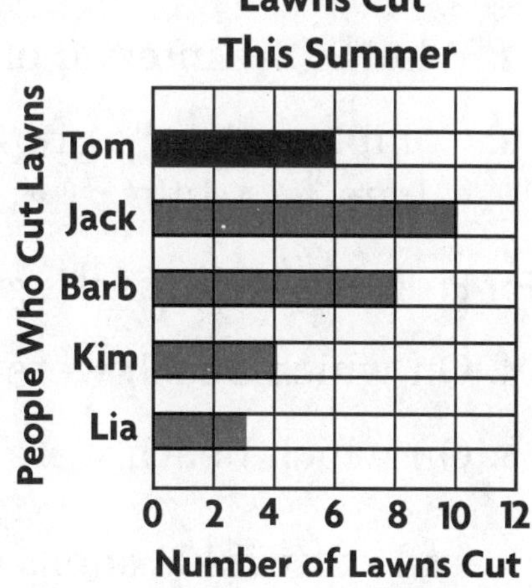

Name ________________________________

Exploring Double-Bar Graphs

Peter made a table of the number of shells that he collected on two beaches. Then he started a **double-bar graph** to compare the data. Each bar represents the number of each kind of shell collected.

SHELLS COLLECTED ON WEST AND DUNE BEACHES

Shell	West Beach	Dune Beach
Clam	6	8
Oyster	9	1
Scallop	10	2
Periwinkle	6	6

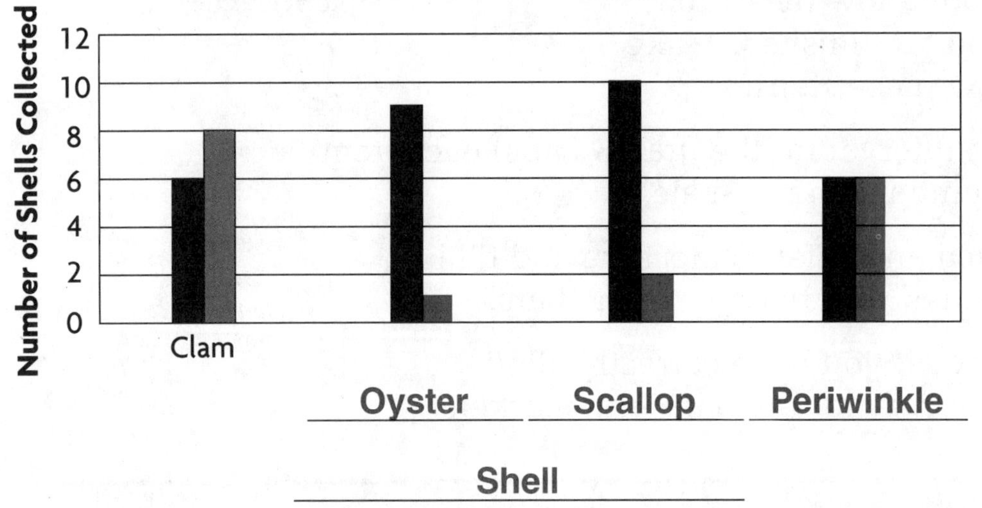

Oyster Scallop Periwinkle

Shell

1. What scale did Peter choose? 0, 2, 4, 6, 8, 10, 12
2. What is the interval of the scale? 2
3. Complete Peter's graph. Remember to include
   - bars.
   - title.
   - labels.

   Check students' graphs.

For Problems 4–5, use the graph.

4. On which beach were more oyster shells collected? West Beach
5. On which beach were an equal number of clam shells and periwinkle shells collected? West Beach

Name ____________________

LESSON 10.2

Reading a Line Graph

Mr. King made this line graph to show the number of bicycles he sold each month for the first six months of the year. This **line graph** shows how bike sales change over time.

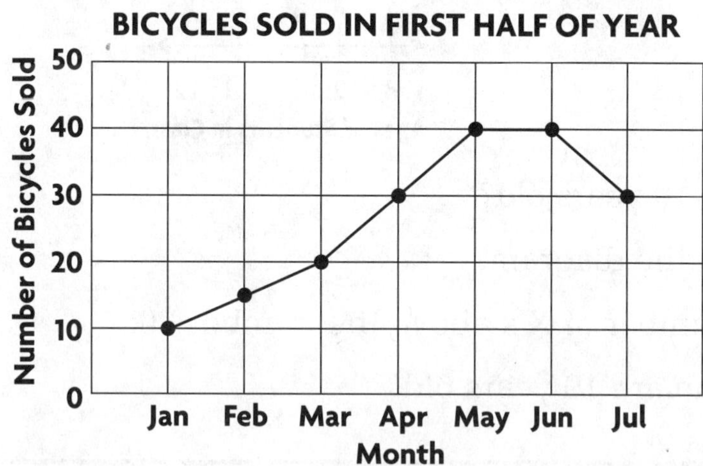

How many bicycles were sold in April?

Step 1 Find the line labeled *April.* Put your finger on the line. Follow that line up to the point (•).

Step 2 Move your finger along the line to the left to locate the number of bicycles sold in April.

Mr. King sold 30 bicycles during April.

For Problems 1–5, use the graph.

1. Mr. King sold the greatest number of bicycles during

   May and **Jun**.

2. During March, Mr. King sold **20** bicycles.

3. During February, Mr. King sold **15** bicycles.

4. Mr. King sold **30** more bicycles in June than in January.

5. There was an increase in the number of bicycles sold from

   Jan to Feb, Feb to Mar, Mar to Apr, and Apr to May

Name ______________________

Line Plot

Ms. Ryan is teaching a cartooning class for students ages 8 to 12. She made a line plot. This **line plot** shows the number of students of each age in the class.

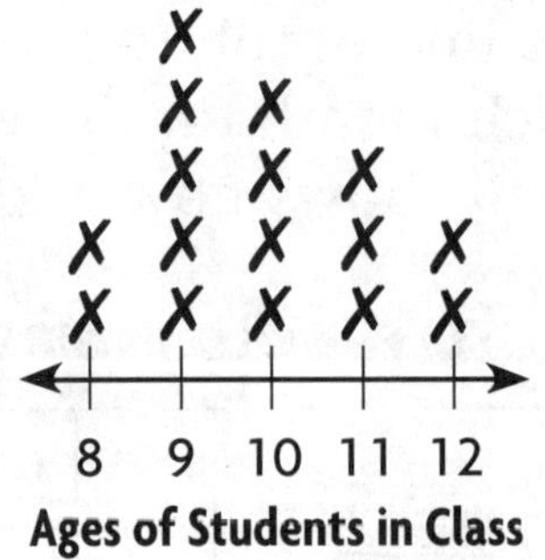

How many students are 10 years old?

Step 1 Locate 10 on the diagram.

Step 2 Count the number of X's above the number 10.

There are 4 students who are 10 years old.

For Problems 1–4, use the line plot.

1. There are ___3___ students who are 11 years old.
2. More students are ___9___ years old than any other age.
3. The same number of students are ___8___ and ___12___ years old.
4. The **range** is the difference between the greatest and least numbers in a set of data. The range in ages of students in the class is ___4 years___.
5. After the first morning of cartooning class, the students counted the number of cartoon pages they had completed. Use the data in the table to make a line plot.

CARTOON PAGES COMPLETED						
Number of Pages	1	2	3	4	5	6
Number of Students	0	2	4	7	2	1

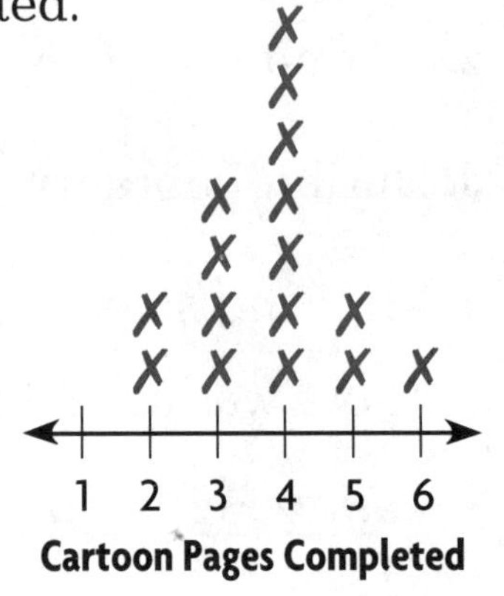

Name ____________________

LESSON 10.4

Stem-and-Leaf Plot

Becky measured the height in inches of each sunflower in her garden. She made a list of the heights.

20 inches	14 inches	17 inches
35 inches	12 inches	15 inches
12 inches	32 inches	27 inches
23 inches	24 inches	21 inches
27 inches	27 inches	33 inches
18 inches	33 inches	

- To organize the data in a **stem-and-leaf plot** write the numbers in order from least to greatest. Group the data in rows according to the digit in the tens place.
- To find the heights of the shortest and tallest sunflowers in the garden, look at the first and last numbers in the stem-and-leaf plot.

Stem (Tens)	Leaves (Ones)
1	2 2 4 5 7 8
2	0 1 3 4 7 7 7
3	2 3 3 5

Heights of Sunflowers in Inches

The shortest sunflower is 12 inches. The tallest sunflower is 35 inches.

For Problems 1–5, use the stem-and-leaf plot.

1. How many sunflowers are 33 inches tall? 2 sunflowers
2. How many sunflowers are 20 or more inches tall? 11 sunflowers
3. The **mode** is the number that occurs most often in a set of data. What is the mode for the sunflower heights? 27 in.
4. The **median** is the middle number in an ordered series of numbers. What is the median sunflower height? 23 in.
5. The range is the difference between the greatest and the least numbers in a data set. What is the range in sunflower heights? 23 in.

Name ______________________

LESSON 10.5

Choosing a Graph

The students in Mr. Schmitt's fourth-grade class participated in a read-a-thon for four weeks. The bar graph shows the number of books of each kind that students read. The line graph shows the number of books students read each week during the read-a-thon.

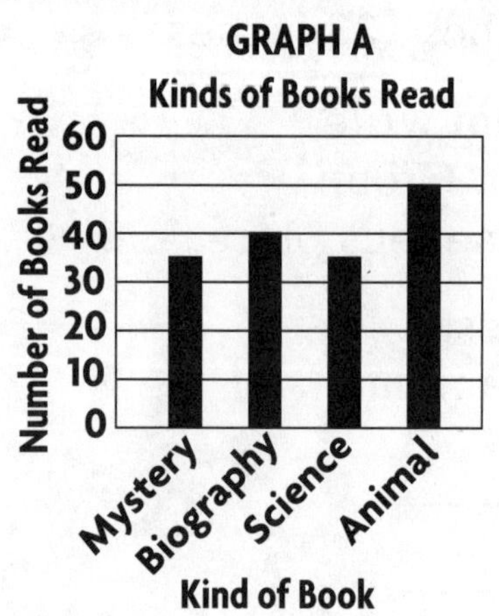

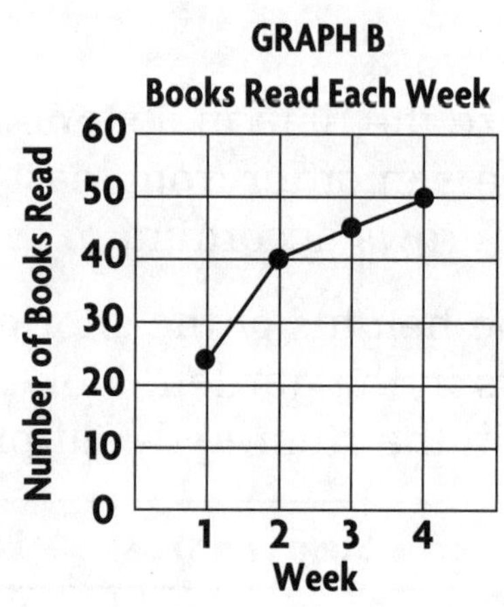

1. What kind of graph is Graph A? **bar graph**

2. What do the bars in Graph A represent? **the number of mystery, biography, science, and animal books read**

3. Did students read more biographies or more science books? How does the graph show this?

   More biographies; the bar for biographies is taller.

4. What kind of graph is Graph B? **line graph**

5. Did the number of books read each week increase or decrease? How does the graph show this?

   Increase; the points and the connecting line go up.

6. What would the line connecting the data points look like if the number of books read each week was the same?

   The line would be straight and horizontal.

7. What other kind of graph could you use to show the data in Graph A? **pictograph**

Name ___________________________________

Problem-Solving Strategy

Make a Graph

During the first week of summer vacation, Andy and Pam opened up a lemonade stand in front of their house. They kept a record of the number of cups of lemonade they sold each afternoon during the week.

Andy and Pam want to show how the sales of lemonade changed during the week.

LEMONADE SALES FOR ONE WEEK	
Day	**Number of Cups Sold**
Monday	8
Tuesday	10
Wednesday	12
Thursday	12
Friday	12
Saturday	20
Sunday	22

Follow these steps to complete the line graph. **Check students' graphs.**

Step 1 Use the table to find how many cups of lemonade were sold on Monday.

Step 2 Find on the graph the line labeled Monday.

Step 3 Follow that line up to the intersecting line labeled 8.

Step 4 Place a point on the graph.

Step 5 Repeat Steps 1–4 for the other six days of the week.

Step 6 Use a straightedge to connect the points on the graph.

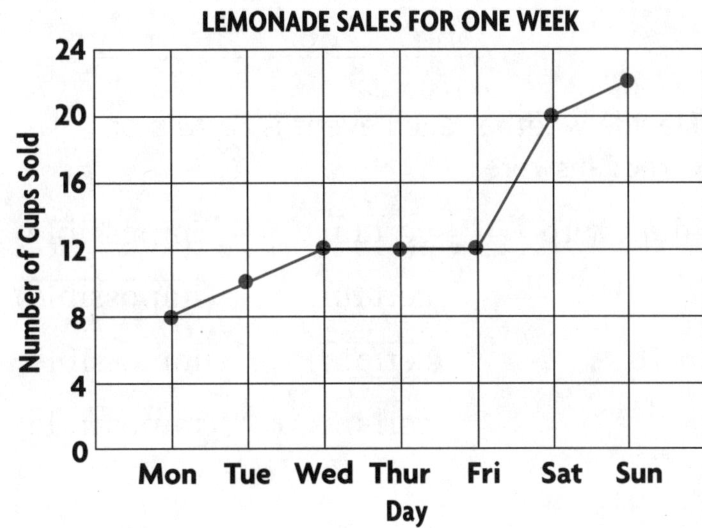

Name ______________________________

Certain and Impossible

Lydia has a number cube with the numbers 1 through 6 on it.

Is rolling a number less than 7 a *certain* event or an *impossible* event?

To find out, list all the numbers Lydia could roll. You can then **predict,** or tell what will happen, when you roll the cube.

- If the event describes all the numbers, it is **certain.**
- If the event describes none of the numbers, it is **impossible.**

Numbers Lydia could roll:	1	2	3	4	5	6
Number less than 7?	yes	yes	yes	yes	yes	yes

The event describes all the numbers. So, it is certain Lydia will roll a number less than 7.

For Problems 1–4, use the numbers below. Decide whether the event describes each number. Write *yes* or *no.*

	1	2	3	4	5	6
1. a number greater than zero	yes	yes	yes	yes	yes	yes
2. a 2-digit number	no	no	no	no	no	no
3. a number less than 10	yes	yes	yes	yes	yes	yes
4. the number 7	no	no	no	no	no	no

Look at Problems 5–8. Decide whether each event is *certain* or *impossible.* Circle the correct answer.

5. a number greater than zero	(certain)	impossible
6. a 2-digit number	certain	(impossible)
7. a number less than 10	(certain)	impossible
8. the number 7	certain	(impossible)

Name ___________________________

LESSON 11.2

Likely and Unlikely

Look at the spinner.

Is spinning a number with 2 as a digit a *likely* or an *unlikely* event?

To find out if an event is *likely* or *unlikely*, list all the numbers you could spin.

- If the event describes half, or more than half, of the numbers, it is *likely*.
- If the event describes fewer than half of the numbers, it is *unlikely*.

Numbers you could spin:	0	6	13	18	24	30
Number with 2 as a digit?	no	no	no	no	yes	no

The event describes fewer than half of the numbers. So, it is *unlikely* you will spin a number with 2 as a digit.

For Problems 1–4, use the spinner above. Decide whether the event describes each number. Write *yes* or *no.*

	0	6	13	18	24	30
1. a number greater than zero	no	yes	yes	yes	yes	yes
2. a 2-digit number	no	no	yes	yes	yes	yes
3. a number less than 10	yes	yes	no	no	no	no
4. a number divisible by 3	no	yes	no	yes	yes	yes

For Problems 5–8, tell whether each event is *likely* or *unlikely*.

5. a number greater than zero ___likely___

6. a 2-digit number ___likely___

7. a number less than 10 ___unlikely___

8. a number divisible by 3 ___likely___

Name ______________________

LESSON 11.3

Predicting Outcomes

When you toss one coin, there are 2 possible ways the coin can land: heads or tails. Each way has the same chance of happening. The chances of heads or tails are equally likely.

When you toss two coins, there are 4 possible ways the coins can land.

There is 1 way to get 2 heads. → (heads, heads)

There are 2 ways to get 1 head and 1 tail. → (heads, tails), (tails, heads)

There is 1 way to get 2 tails. → (tails, tails)

POSSIBLE WAYS FOR TWO COINS TO LAND	
First Coin	**Second Coin**
heads	heads
heads	tails
tails	heads
tails	tails

So, tossing 1 head and 1 tail is a likely outcome.

When you spin once, there are 3 possible numbers you can get: 2, 3, or 4.

Complete the table to show all the possible sums you can get if you spin twice and add the numbers.

	POSSIBLE SUMS FOR TWO SPINS		
	First Spin	**Second Spin**	**Sum**
1.	2	2	2 + 2 = 4
2.	2	3	2 + 3 = 5
3.	2	4	2 + 4 = 6
4.	3	2	3 + 2 = 5
5.	3	3	3 + 3 = 6
6.	3	4	3 + 4 = 7
7.	4	2	4 + 2 = 6
8.	4	3	4 + 3 = 7
9.	4	4	4 + 4 = 8

10. Look at the sums. Which sum is most likely to happen? 6

Name ___________________________________

Probability

Probability can be described by a number written as a fraction.

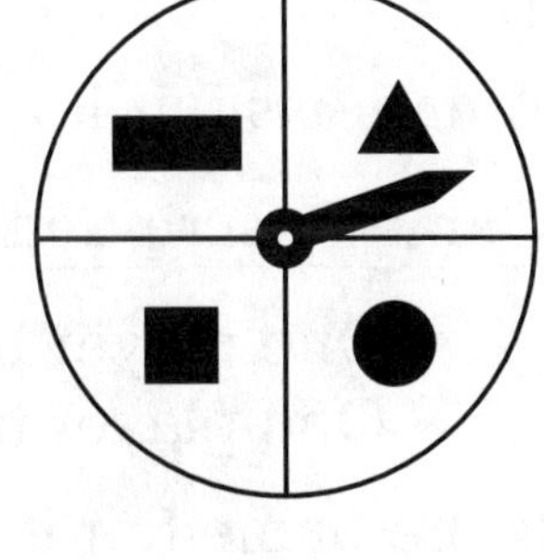

$$\text{Probability} = \frac{\text{number of outcomes the event describes}}{\text{number of outcomes possible}}$$

Using the spinner at the right, what is the probability of spinning a circle?

- List all the outcomes possible.

4 outcomes

- Identify the outcomes that are circles.

1 circle

- Record the probability as a fraction.

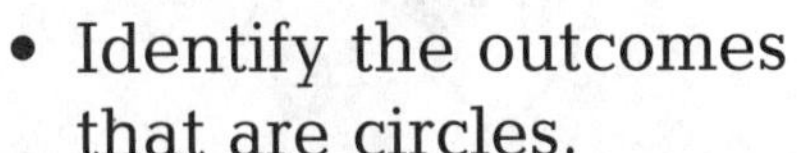
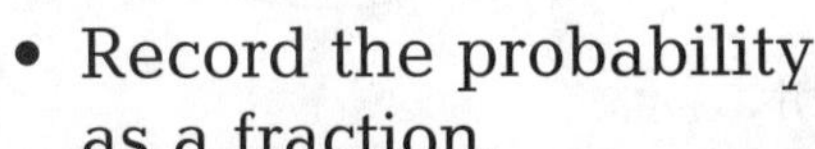

$$\text{Probability circle} = \frac{\text{number of circles}}{\text{number of shapes}} = \frac{1}{4}$$

Use the spinner at the right. Circle the outcomes the event describes. Then write the probability as a fraction. The first one has been done for you.

1. What is the probability of spinning a ball?

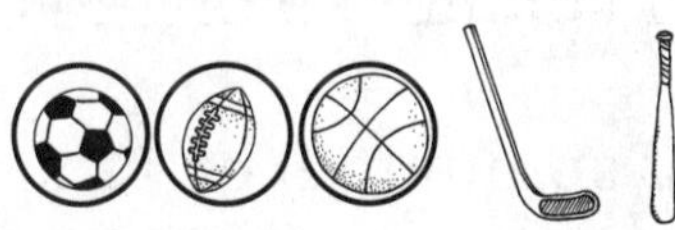

$\frac{3}{5}$

2. What is the probability of spinning a ball that is round?

$\frac{2}{5}$

3. What is the probability of spinning something that is not a ball?

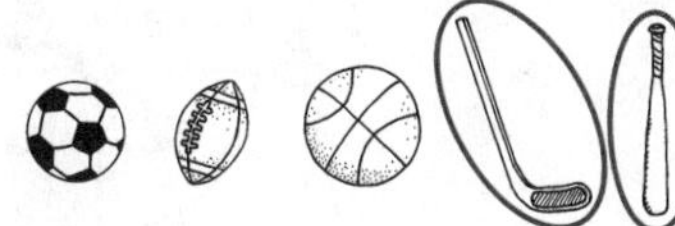

$\frac{2}{5}$

4. What is the probability of spinning a football?

$\frac{1}{5}$

5. What is the probability of spinning a ball or a baseball bat?

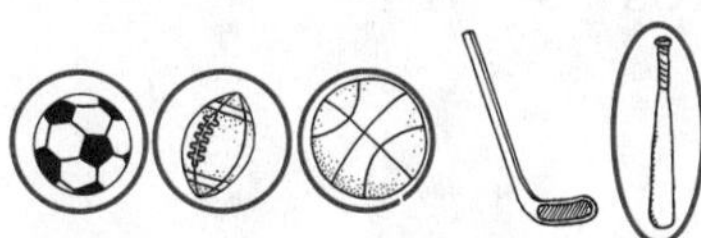

$\frac{4}{5}$

Name ______________________

LESSON 11.5

Testing for Fairness

Fairness of a game is when one player is not more likely to win than another. To see if a game is fair, compare the probabilities of winning and not winning. A game is fair if the probabilities are equal.

Game Description

- One player tries to spin even numbers.
- One player tries to spin odd numbers.

Is the game fair if this spinner is used?

probability of spinning an even number $= \frac{3}{5}$
probability of spinning an odd number $= \frac{2}{5}$

The probabilities are not equal, so the game is not fair.

Use the spinner above. Write the probabilities. Write whether the game is *fair* or *not fair*.

Game Description

- One player tries to spin 9 or 12.
- One player tries to spin 8 or 12.

1. The probability of spinning 9 or 12 is $\frac{2}{5}$.

2. The probability of spinning 8 or 12 is $\frac{2}{5}$.

3. The game is fair.

Write the letter of the spinner that makes the game fair.

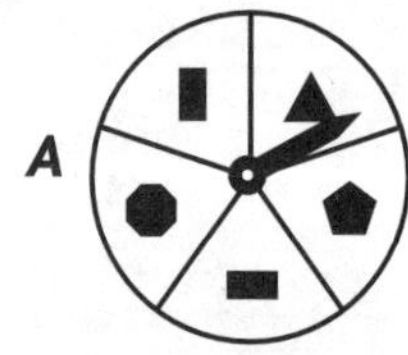

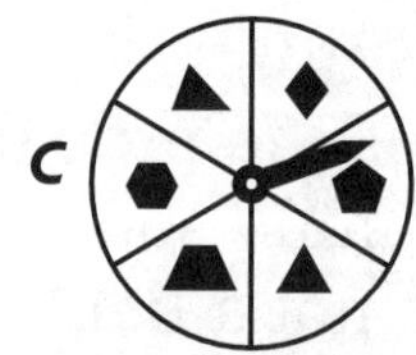

4. Game Description
   - One player tries to spin a rectangle.
   - One player tries to spin a triangle.

   Spinner B

5. Game Description
   - One player tries to spin a shape with 4 or fewer sides.
   - One player tries to spin a shape with 4 or more sides.

   Spinner C

Name ________________________________

Problem-Solving Strategy

Make a Model

Daryl, Tat, Judy, and Nora have written their first initials in sections of a spinner to play a game. The spinner has 8 equal sections. In 3 of the sections is a *D*. The sections with a *T* take up $\frac{1}{4}$ of the spinner. There is the same number of sections with *J* as with *T*. Each remaining section has an *N*. Is the spinner fair?

Make your model step by step, using one fact at a time.

- The spinner has 8 equal sections.

- In 3 of the sections is a *D*.

- The sections with a *T* take up $\frac{1}{4}$ of the spinner.

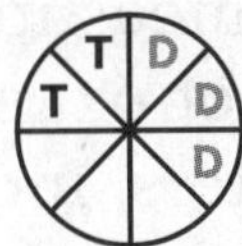

- There is the same number of sections with *J* as with *T*.

- Each remaining section has an *N*.

- Since Daryl has a greater chance of spinning his initial, *D*, the spinner is not fair.

Read the problem. Complete each fact and then complete the model to solve. Use crayons.

Kelsey made a spinner with 10 equal sections. She colored 2 opposite sections blue. She colored the section before each blue section orange. She made $\frac{2}{10}$ of the spinner green and $\frac{2}{10}$ of the spinner red. The rest of the sections she colored purple. Did Kelsey make a fair spinner?

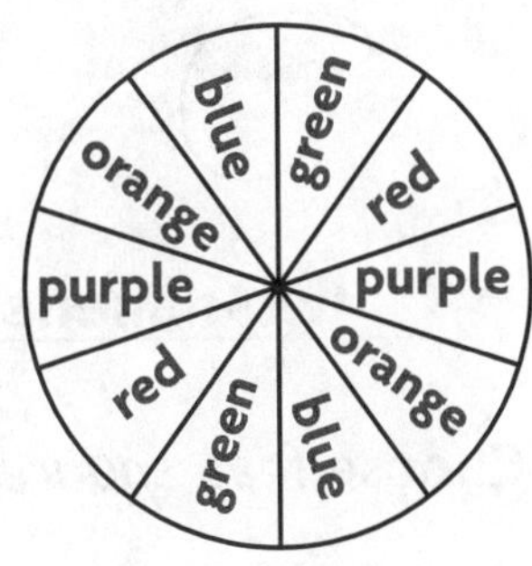

1. How many sections are each color?

   A. blue **2 sections** B. orange **2 sections** C. green **2 sections**

   D. red **2 sections** E. purple **2 sections**

2. The spinner **is** fair.

Name ______________________________

Exploring Geometric Figures

Geometric figures can be one-dimensional, two-dimensional, or three-dimensional. Figures such as lines, rectangles, and solids have **dimension** and can, therefore, be measured.

Lines, rays, and line segments are one-dimensional figures.

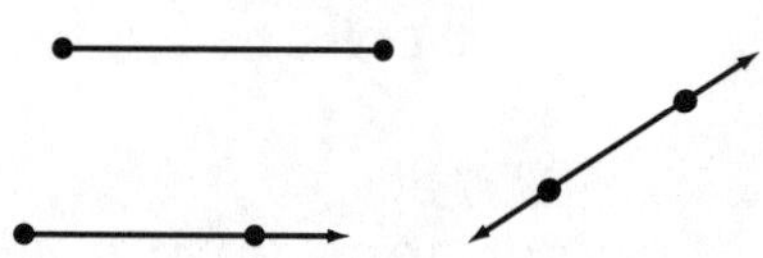

You can use linear units to measure the length (for example, feet, meters, inches, centimeters).

Circles, polygons, and planes are two-dimensional figures.

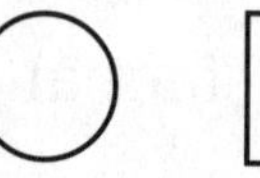

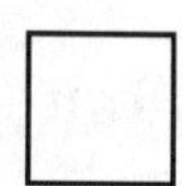

You can use square units to measure the area (for example, square inches, square meters).

Solid figures are three-dimensional.

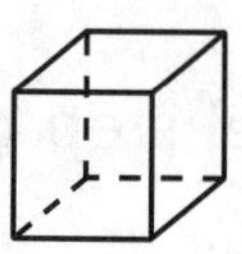

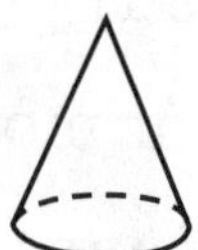

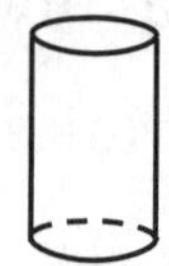

You can use cubic units to measure the volume (for example, cubic inches, cubic yards).

For each example, write *one-dimensional, two-dimensional,* or *three-dimensional* to describe each figure.

1. 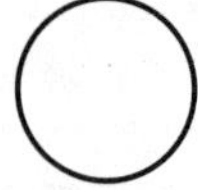

two-dimensional

2. 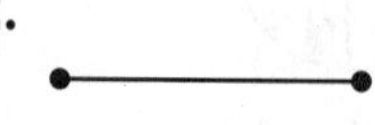

one-dimensional

3. 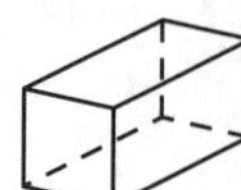

three-dimensional

4. 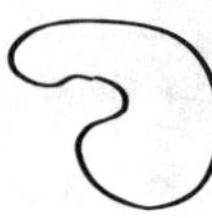

two-dimensional

5. 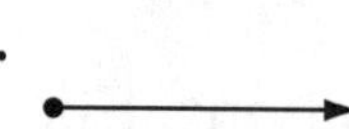

one-dimensional

6. 

one-dimensional

Choose *feet, square feet,* or *cubic feet* to measure each.

7. 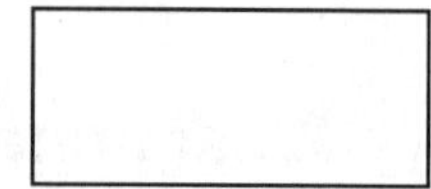

sq ft

8. 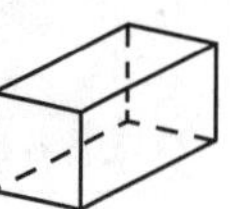

cubic ft

9. 

ft

Name ______________________________

Patterns for Solid Figures

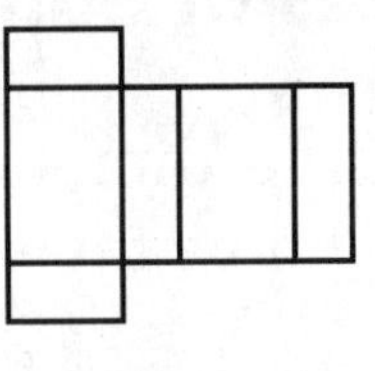

You can make a three-dimensional figure from a two-dimensional pattern called a **net.** Look at the net at the right. Can you see how it forms the three-dimensional rectangular prism?

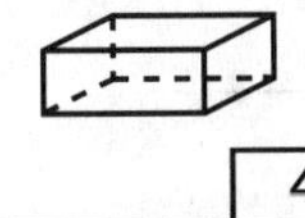

Take another look at the net, and count the faces.

Since the three-dimensional rectangular prism has 6 faces, the net must also have 6 faces.

1 2 3 4 5 6

Draw a line from the three-dimensional figure to its net.

1. 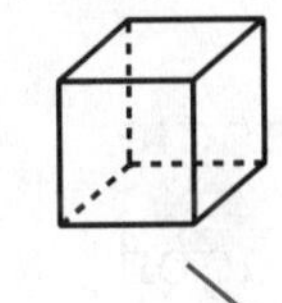

2. 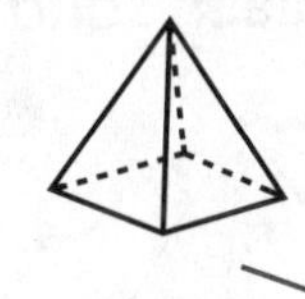

3. 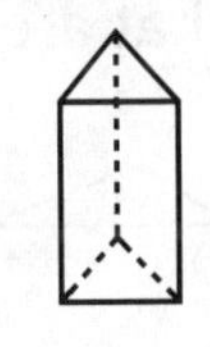

4. 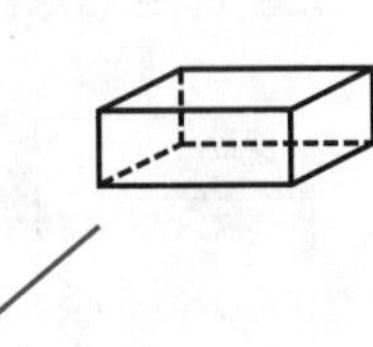

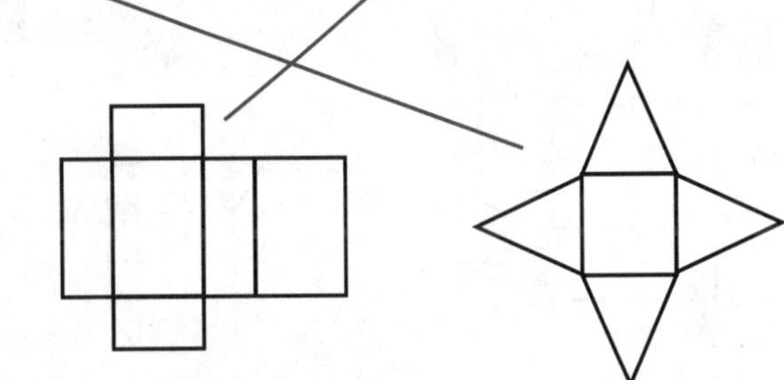

5. 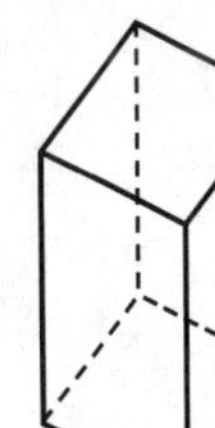

6. 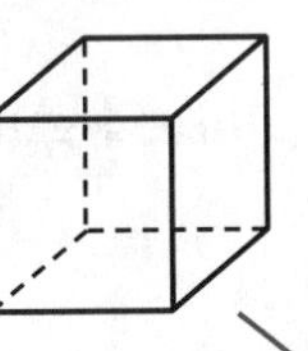

7. 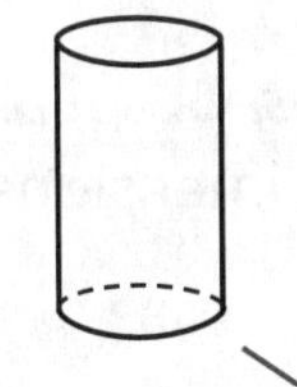

8. 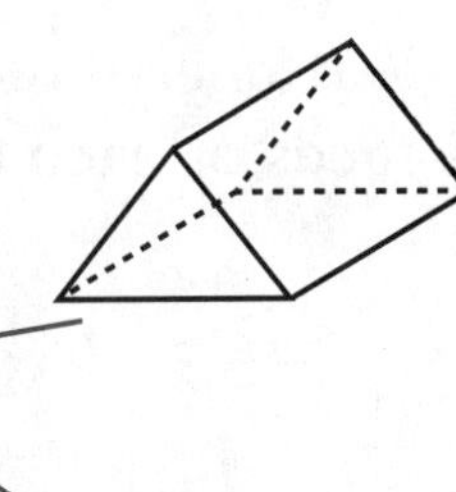

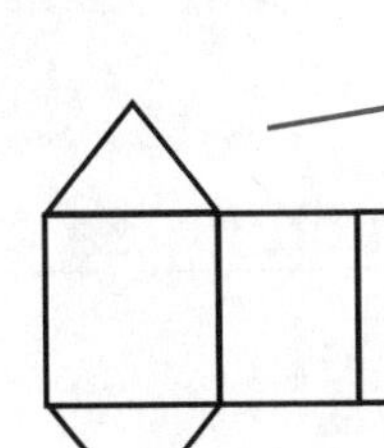
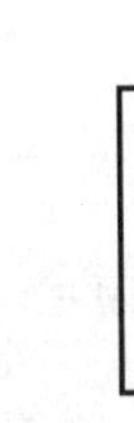
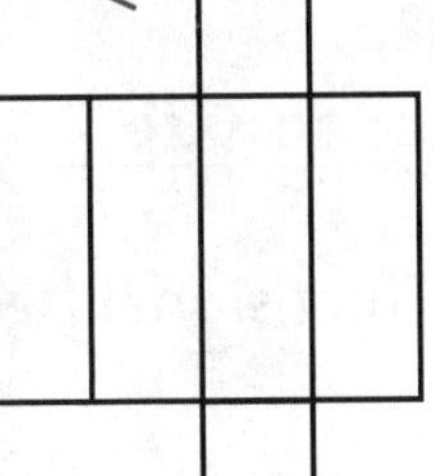
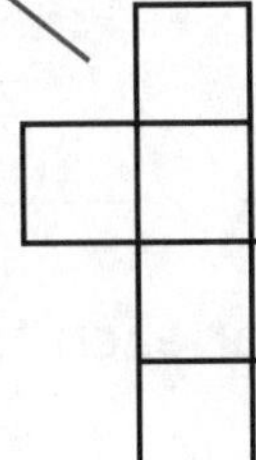
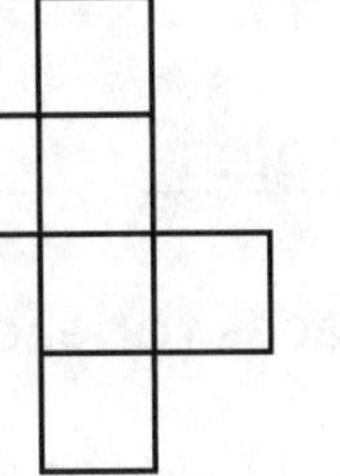
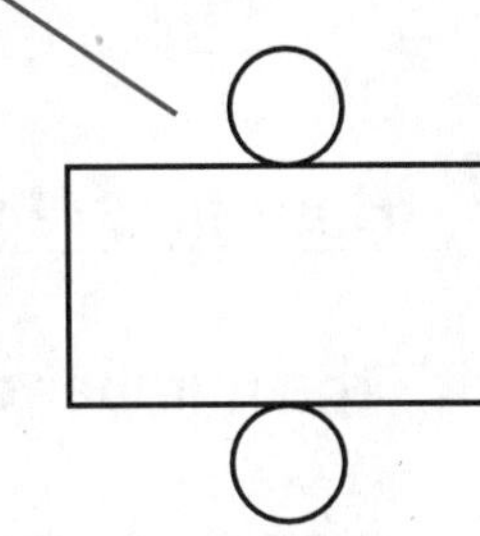

Name ____________________

Faces of Solid Figures

A three-dimensional figure is made up of plane figures, known as faces.

A cube has 6 faces.

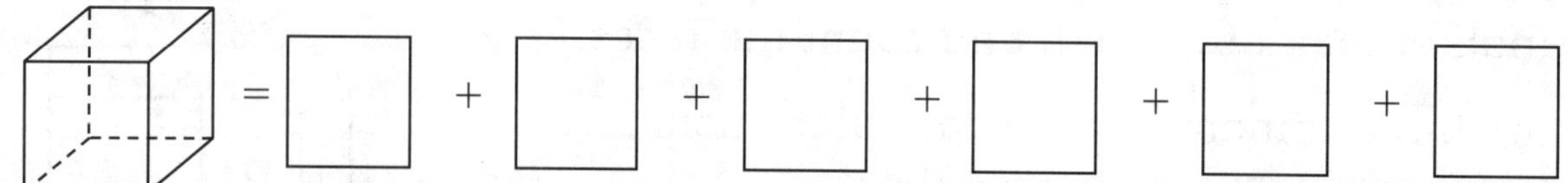

Cutting the cube apart gives you 6 plane figures, or 6 squares.

Some plane figures are triangles, rectangles, and squares.

Choose the plane figures that are the faces of each solid figure.

a. rectangle
b. pentagon
c. triangle
d. square

1. 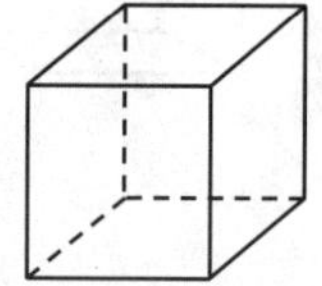 d

2. 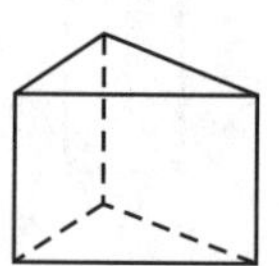 a and c

3. 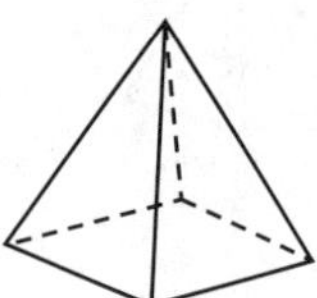 c and d

4. 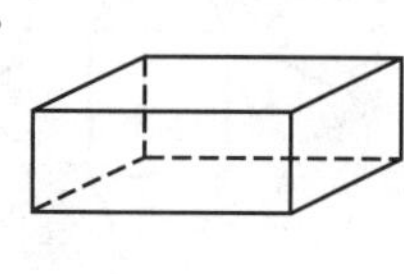 a

5. 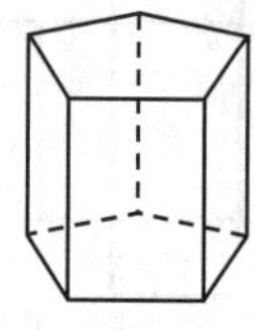 a and b

6. 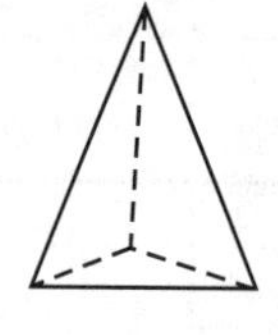 c

Write *triangles, rectangles,* or *squares* for the plane figures that are the faces of each three-dimensional figure.

7. 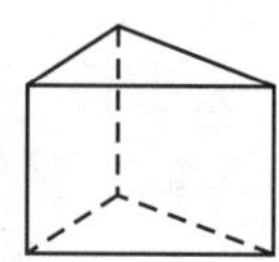 triangles; rectangles

8. 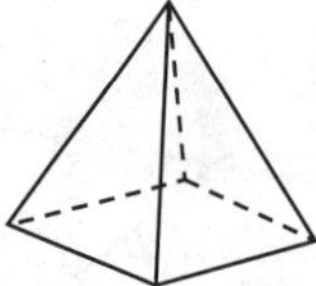 triangles; square

9. 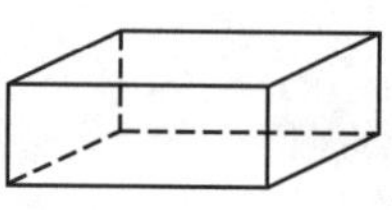 rectangles

Write the number of faces for each three-dimensional figure.

10. rectangular prism 6 faces

11. square pyramid 5 faces

Name ______________________

LESSON 12.3

More About Solid Figures

Three-dimensional figures are made up of faces.

- When two faces meet, they form a straight line, called an **edge.**
- A **vertex** is the point where three or more edges meet in a solid figure. **Vertices** is the word you use to describe more than one vertex.

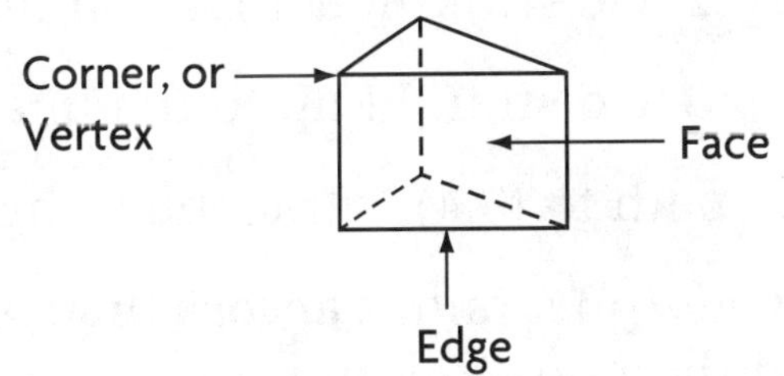

Circle each vertex and write the number of vertices.

1. 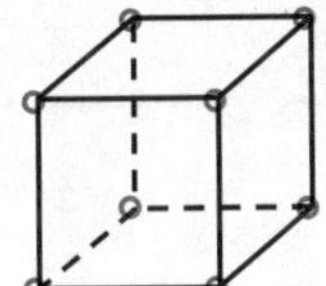
8 vertices

2. 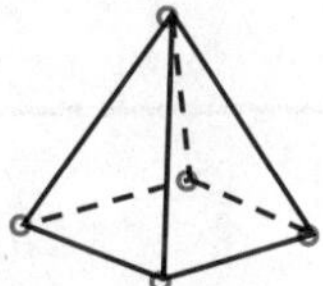
5 vertices

3. 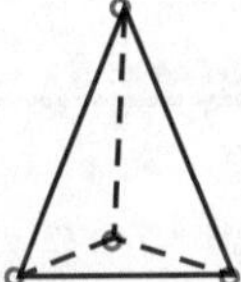
4 vertices

Write the number of edges for each three-dimensional figure.

4. 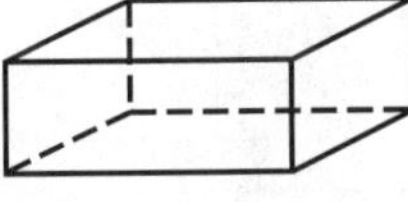
12 edges

5. 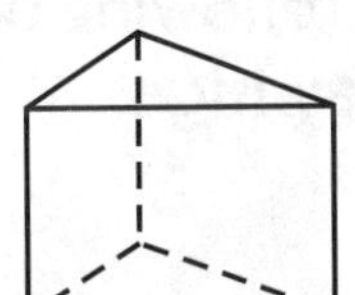
9 edges

6. 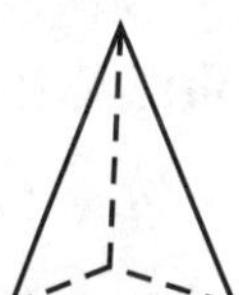
6 edges

Write the number of faces for each three-dimensional figure.

7. 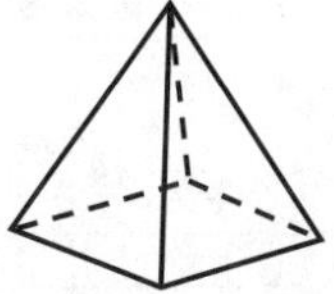
5 faces

8. 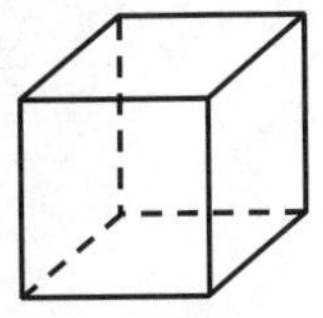
6 faces

9. 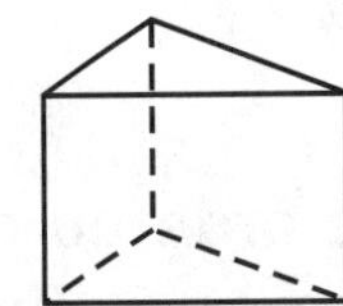
5 faces

Name ______________________

LESSON 12.4

Plane Figures on a Coordinate Grid

You can find the point (3,4) on the grid.

Step 1 Start at 0.

Step 2 Go straight across to number 3.

Step 3 Go straight up to the line labeled 4.

The path to (3,4) is traced on the grid.

Always go straight **across** first, and then go straight **up.**

Remember: Across begins with an **A,** and **A** comes first in the alphabet.

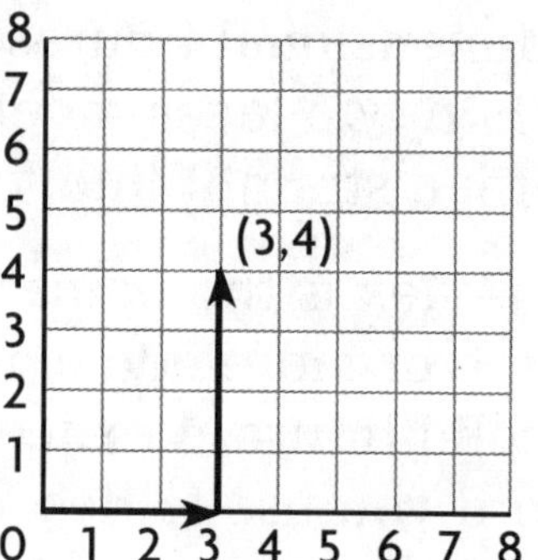

Complete.

1. The numbers in an ordered pair represent the number of spaces you move across and up to locate a point on the grid.

Trace the path from zero to each of the following points on the grid. Place a point. Check students' drawings.

2. (3,4) 3. (5,6)

4. (2,3) 5. (6,7)

6. (4,5) 7. (7,8)

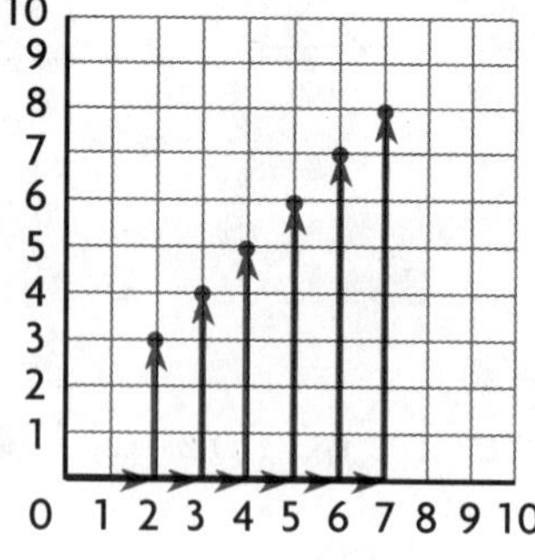

8. Mark the ordered pairs on the grid. Place a point. Connect the points to make a figure.

(2,3), (4,1), (7,1), (9,3), (7,5), (4,5)

9. Name the figure you drew. hexagon

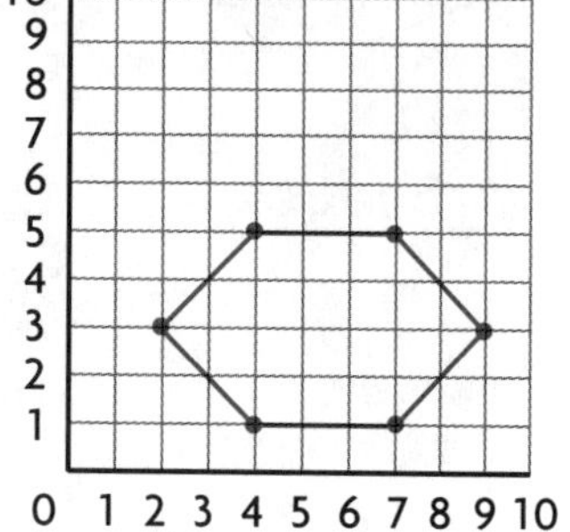

Name ______________________

Classifying and Sorting Solid Figures

There are many kinds of solid figures. They can have flat faces, curved surfaces, or both flat faces and curved surfaces.

cylinder
2 flat faces
1 curved surface

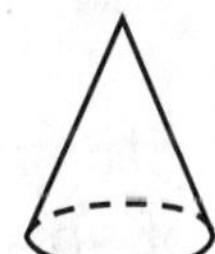

cone
1 flat face
1 curved surface

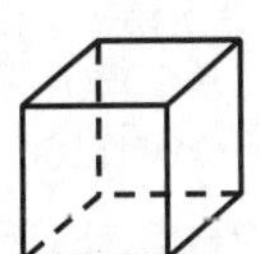

cube
6 flat faces
0 curved surfaces

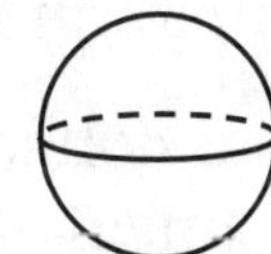

sphere
0 flat faces
1 curved surface

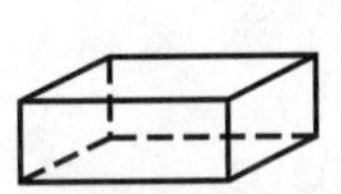

rectangular prism
6 flat faces
0 curved surfaces

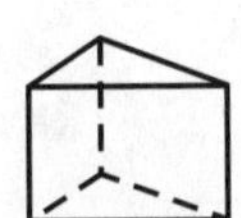

triangular prism
5 flat faces
0 curved surfaces

triangular pyramid
4 flat faces
0 curved surfaces

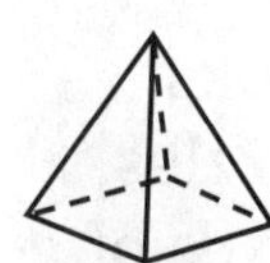

square pyramid
5 flat faces
0 curved surfaces

Write the names of the solid figures that answer each question.

1. Which solid figures have both flat faces and curved surfaces?

   cylinder, cone

2. Which solid figures have a square face?

   cube, square pyramid

3. Which solid figures have a curved surface?

   cylinder, cone, sphere

4. Which solid figures have no curved surfaces?

   cube, rectangular prism, triangular prism, triangular pyramid, square pyramid

5. Which solid figures have fewer than 5 flat faces?

   cylinder, cone, sphere, triangular pyramid

Name ____________________

LESSON 12.5

Problem-Solving Strategy

Make an Organized List

You can classify and sort things in many ways. One strategy is to *Make an Organized List.*

A school collected the following items for charities: books, puzzles, paper towels, shampoo, cans of soup, boxes of cornflakes, baseballs, oranges, notebooks, colored markers.

You can organize these items by geometric shape.

Box-Shaped	**Sphere-Shaped**	**Cylinder-Shaped**
books	baseballs	paper towels
puzzles	oranges	shampoo
boxes of cornflakes		cans of soup
notebooks		colored markers

Make an organized list to solve. **Possible answers are given.**

1. In Brigitte's locker there are books, coloring pens, a radio, a pencil, a lunch box, chalk, a tennis ball, and a pack of writing paper. Sort the objects by shape.

Box-Shaped	**Cylinder-Shaped**	**Sphere-Shaped**
books	coloring pens	tennis ball
radio	pencil	
lunch box	chalk	
pack of paper		

2. At the grocery store, Roger buys milk, butter, juice boxes, cereal, breadsticks, peanut butter, and an orange. Sort the objects by shape.

Box-Shaped	**Cylinder-Shaped**	**Sphere-Shaped**
milk, cereal	breadsticks	orange
butter, juice boxes	peanut butter	

Name ______________________________

Line Segments

Learning some basic terms can help you understand geometry.

Points identify locations on objects and in space. Points are named by letters. This is point *A*.

A **plane** is a flat surface with no end. Planes are named by any three points in the plane. This is plane *ABC*.

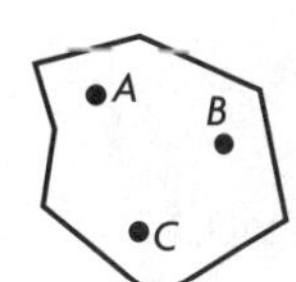

A **line** is a straight path in a plane. It has no end. It can be named by any two points on the line. This is line *AB*.

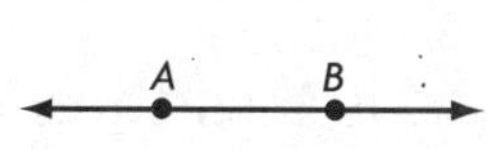

A **line segment** is the distance between two points on a line. This is line segment *AB*.

Parallel line segments never cross. They are always the same distance apart. Line segment *AB* is parallel to line segment *CD*.

Match the term with the drawing.

1. line __C__ **2.** points __B__ **3.** line segment __D__ **4.** plane __A__

A.

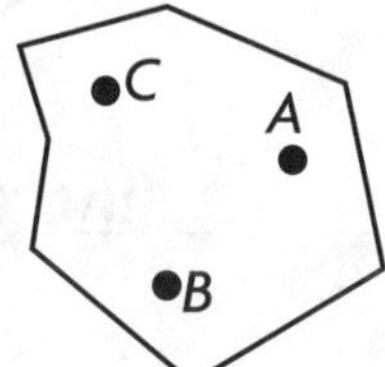

B.

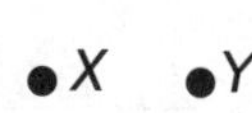

C.

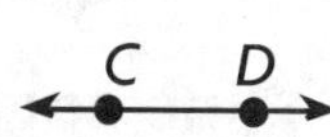

D.

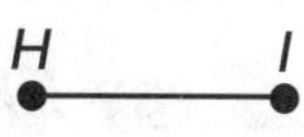

Write *points, line,* or *line segment* to name each figure.

5. A B

__line segment__

6. G H

__points__

7.

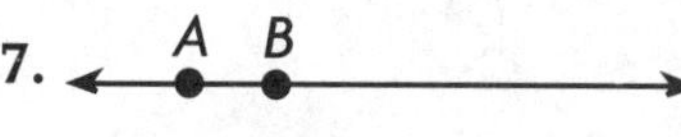

__line__

Circle the parallel line segments.

8.

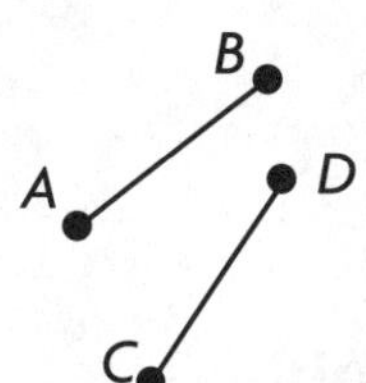

F H I G

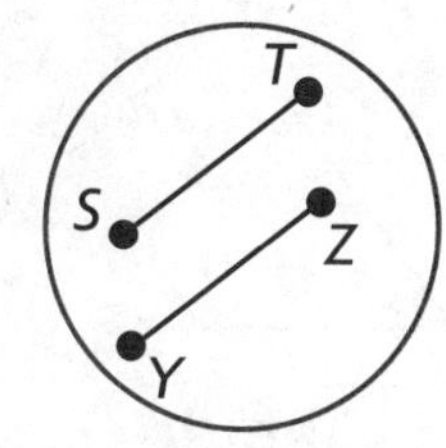

Name ______________________________

LESSON 13.2

Exploring Angles and Line Relationships

A **ray** is part of a line and has one endpoint. You may think of the sun's rays, which start at the sun (endpoint) and go on forever into space.

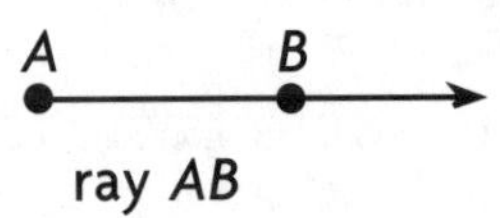

ray *AB*

An **angle** is formed when two rays have the same endpoint.

D E C

angle C of *ECD*

There are three types of angles:

A **right angle** forms a square corner.

An **acute angle** is less than a right angle.

An **obtuse angle** is greater than a right angle.

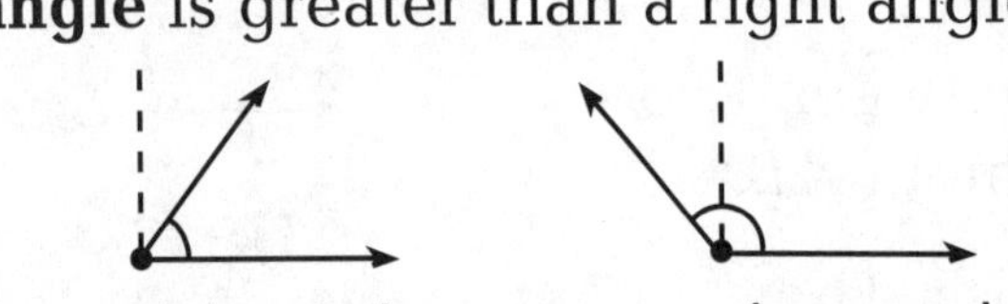

Hint: *Obtuse* begins with an **O**. Think of an **O**pen angle.

Write the name of each figure.

1. E F

ray *EF*

2. Y Z

line segment *YZ*

3. B

point *B*

4. C D

line *CD*

Write *acute, right,* or *obtuse* to name each angle.

5. 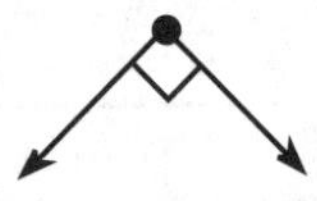

right

6. 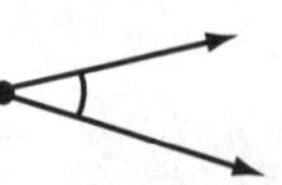

acute

7. 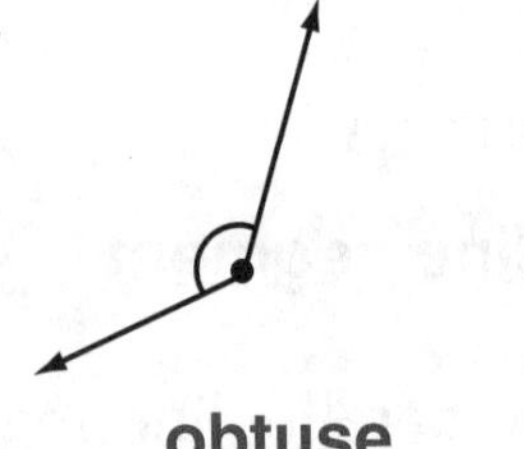

obtuse

8. 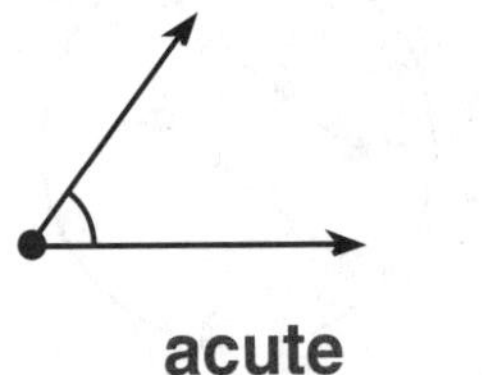

acute

9. 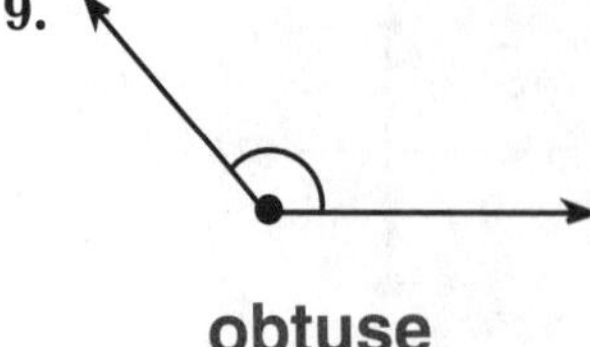

obtuse

10. 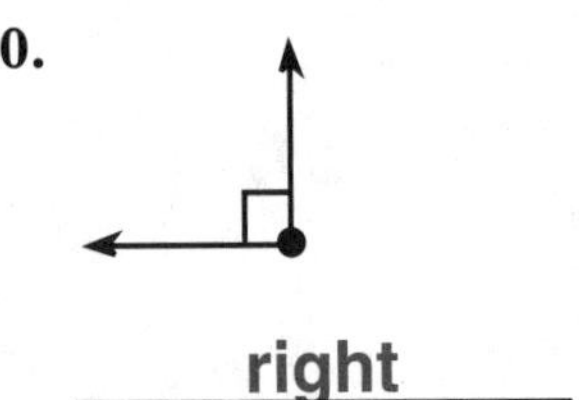

right

Name ______________________________

More About Angles and Line Relationships

All lines that cross each other are **intersecting lines.** Lines *DE* and *FG* intersect.

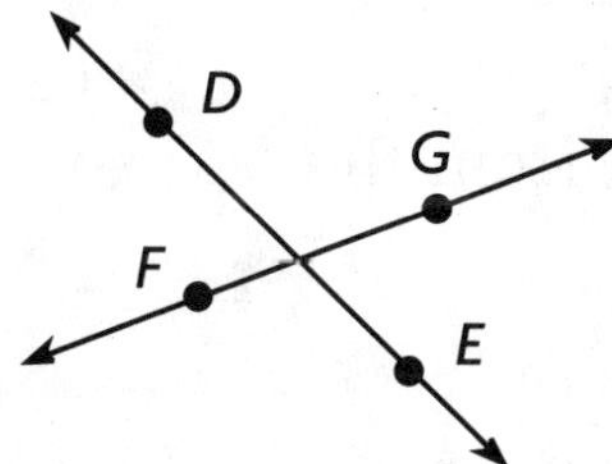

Lines that intersect to form four right angles are **perpendicular lines.** Lines *AB* and *CD* are perpendicular lines.

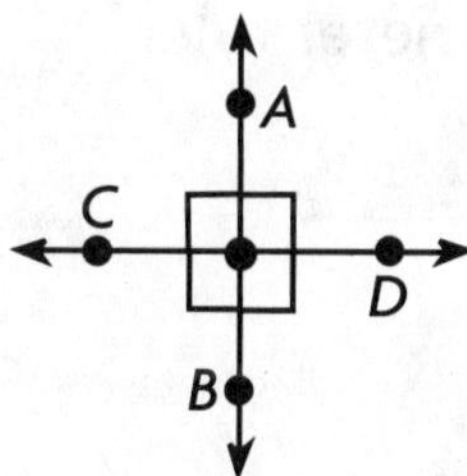

In Exercises 1–3, draw perpendicular lines. Check students' drawings.

1. 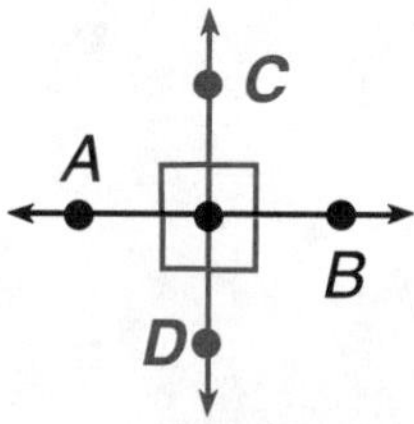

2. 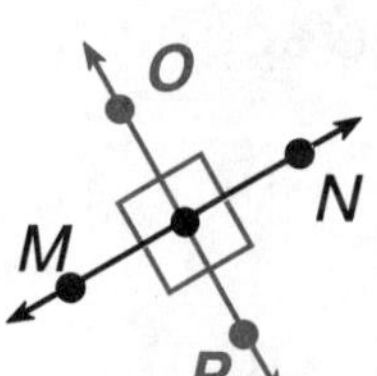

3. 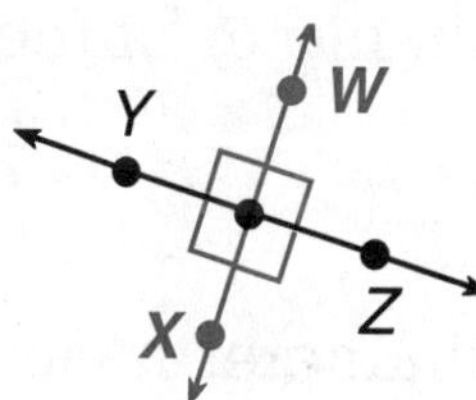

For Exercises 4–5, use the following map.

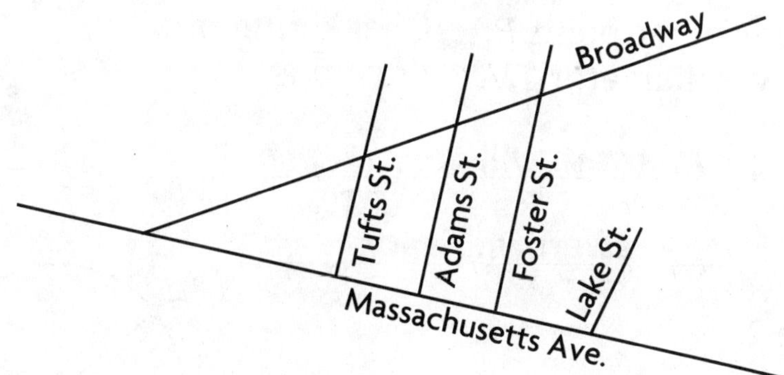

4. Name the streets parallel to Adams Street. Tufts St., Foster St.

5. Name the streets perpendicular to Massachusetts Avenue.

Tufts St., Adams St., Foster St.

Name ______________________________

Exploring Circles

A **circle** is named by its center point. All the points on the circle are the same distance from its center point.

A **radius** of a circle is a line segment from the center to any point on the circle.

A **diameter** of a circle is a line segment that passes through the center and has its endpoints on the circle.

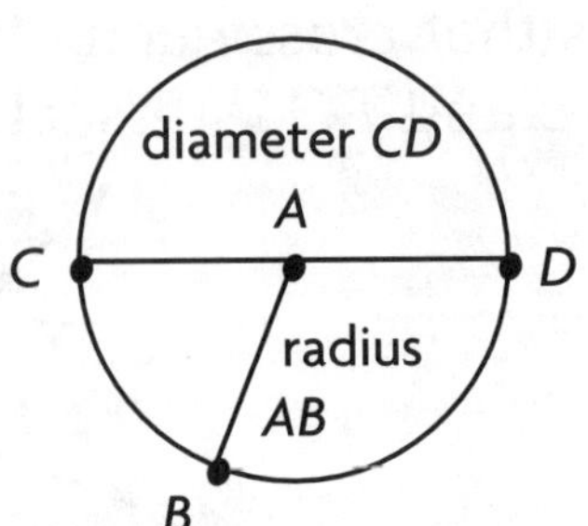

For Problems 1–4, use Circle 1 and a centimeter ruler.

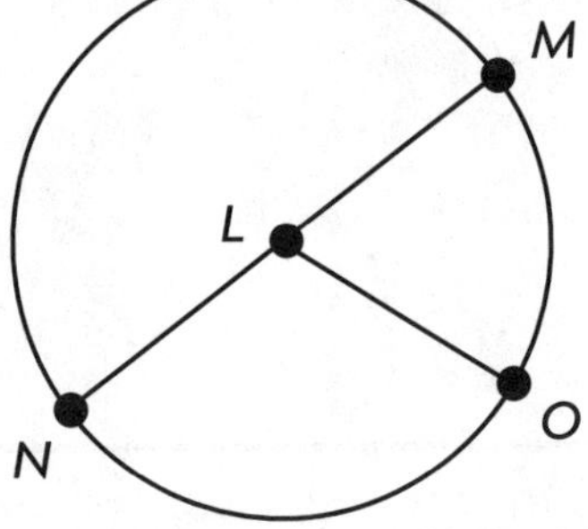

1. The center of the circle is point __L__.

2. Name each radius of the circle.

   __LO, LM, LN__

   Each radius measures __2__ cm.

3. The diameter of the circle is line segment

   __MN__.

   The diameter measures __4__ cm.

4. Name three points on the circle. __M, N, O__

Circle 1

5. Draw a circle. Label the center point *R*. Draw a radius *RS*. Draw a diameter *TV*. **Check students' drawings.**

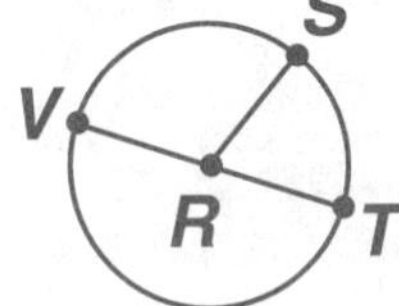

   Measure the radius. __Answers will vary.__

   Measure the diameter. __Answers will vary.__

For Problems 6–8, use Circle 2.

6. Name the center of the circle. __E__

7. Name each radius of the circle. __EH, EF, EG__

8. Name the diameter of the circle. __HG__

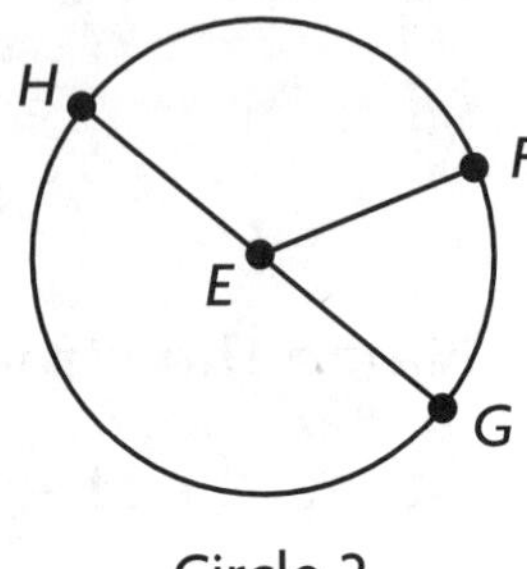

Circle 2

Name ______________________________

LESSON 13.4

Polygons

A **polygon** is a closed plane figure with straight sides. A polygon is named by its number of sides or angles.

A **triangle** has three sides and three angles. The prefix *tri-* has three letters.

A **quadrilateral** has four sides and four angles. The prefix *quad-* has four letters.

A **pentagon** has five sides and five angles. The prefix *penta-* has five letters.

A **hexagon** has six sides and six angles. The prefix *hex-* has an x like *six*.

An **octagon** has eight sides and eight angles. Remember that it is similar to an eight-armed *oct*opus.

Write *yes* or *no* to tell if the figure is a polygon.

1. yes
2. no
3. yes
4. no
5. no
6. yes
7. yes
8. no

Name each polygon. Tell how many sides and angles.

9. quadrilateral, 4 sides, 4 angles
10. hexagon, 6 sides, 6 angles
11. octagon, 8 sides, 8 angles
12. triangle, 3 sides, 3 angles

Use a straightedge to draw each figure. **Possible drawings are shown.**

13. pentagon

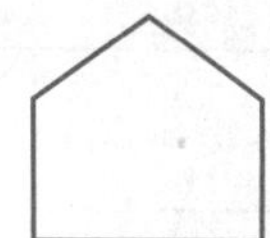

14. hexagon

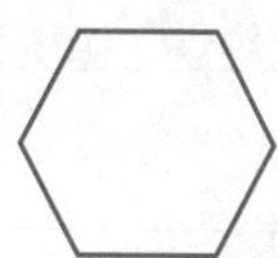

Name ______________________

LESSON 13.5

Quadrilaterals

A **quadrilateral** has four sides and four angles. Quadrilaterals are grouped by their angles, by their sides, and by any parallel sides. The table will help you compare different quadrilaterals.

Grouping	Side Lengths	Parallel Sides	Angles
General Quadrilaterals	4 different lengths	sides may or may not be parallel	angles vary
Trapezoids	lengths can vary	1 pair parallel sides	angles vary
Parallelograms	2 pairs equal lengths	2 pairs parallel sides	2 acute, 2 obtuse, or 4 right
Rectangles	2 pairs equal lengths	2 pairs parallel sides	4 right
Rhombuses	4 equal lengths	2 pairs parallel sides	2 acute, 2 obtuse
Squares	4 equal lengths	2 pairs parallel sides	4 right

Match the drawing with the term.

1. E
2. A
3. F
4. C
5. B
6. D

A. rhombus
B. trapezoid
C. square
D. rectangle
E. parallelogram
F. general quadrilateral

For Exercises 7–8, use the table above.

7. How are rhombuses and squares alike?

They have equal sides and 2 pairs of parallel sides.

8. How are rhombuses and squares different?

A square has 4 right angles; a rhombus has no right angles.

Name ____________________

Problem-Solving Strategy

Act It Out

You can solve a problem involving geometric figures by *acting it out.*

Trace the triangles at the right and cut them out.

For Exercises 1–4, use the four triangles to make different quadrilaterals. Draw a picture of each quadrilateral. **Check students' drawings.**

1. Draw a square.

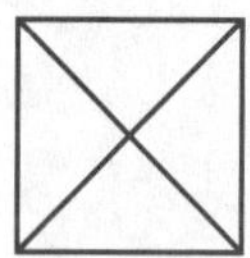

2. Draw a rectangle.

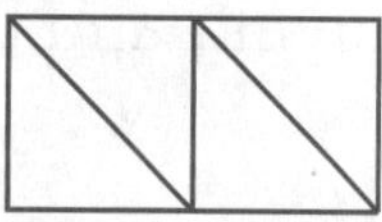

3. Draw a trapezoid.

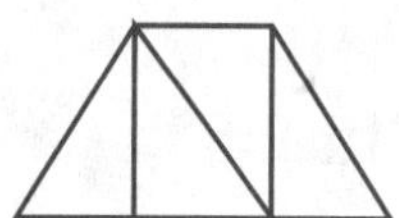

4. Draw a parallelogram.

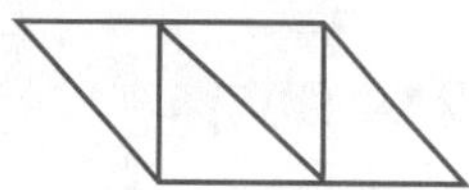

Solve.

5. Lockers are grouped in fours at the Melick School. Ariana's locker is after Hasan's but before George's. Eraldo's locker is after Hasan's but before Ariana's. If Ariana's locker is next to George's, in what order are the lockers?

   Hasan's, Eraldo's, Ariana's, George's

6. The school store sold T-shirts at the school fair. In the morning 134 shirts were sold. In the afternoon 187 shirts were sold. The store had 29 shirts remaining. How many shirts did the store have at the beginning of the day?

   350 shirts

Name ____________________

LESSON 14.1

Finding Perimeter

Perimeter is the distance around a figure. You can find the perimeter by adding the lengths of the sides of a figure.

7 in.
3 in. 2 in. 3 in.
1 in.
10 in.

The perimeter is
6 + 4 + 6 + 4 = 20 units.

The perimeter is
1 in. + 3 in. + 2 in. + 7 in. + 3 in. + 10 in. = 26 in.

If two pairs of sides are the same length, you can find the perimeter by multiplying and adding.

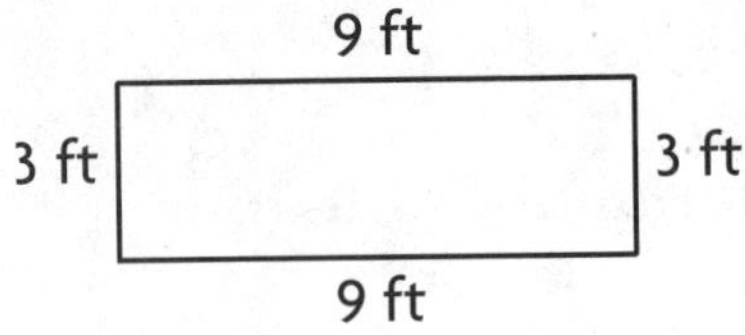

10 in.
2 in. 2 in.
10 in.

(2 × 3 ft) + (2 × 9 ft) =
6 ft + 18 ft = 24 ft

(2 × 2 in.) + (2 × 10 in.) =
4 in. + 20 in. = 24 in.

Find the perimeter.

1. 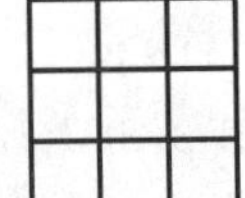
12 units

2. 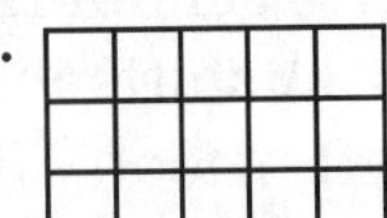
16 units

3. 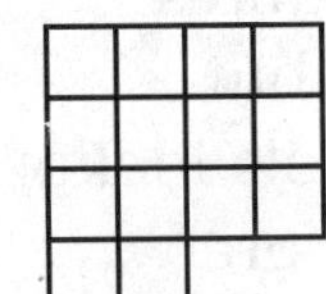
16 units

4. 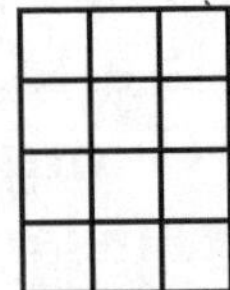
14 units

5. 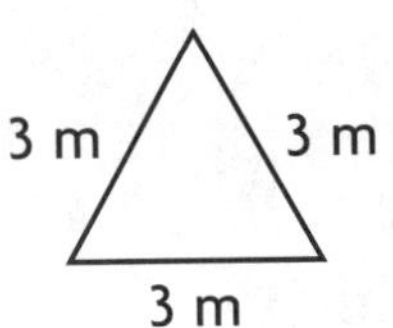

9 m

6. 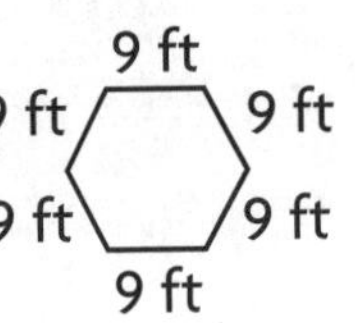

54 ft

7. 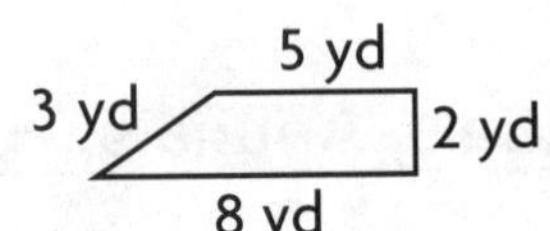

18 yd

8. 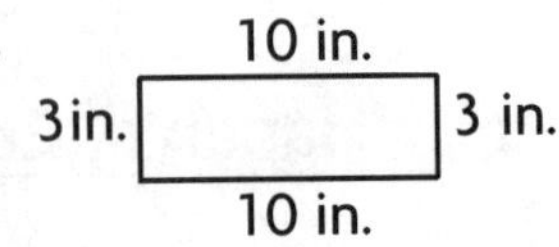

26 in.

Name ______________________________

Exploring Area

Area is the number of square units needed to cover a flat surface.

You can find the area of each figure on the geoboard by counting the number of square units.

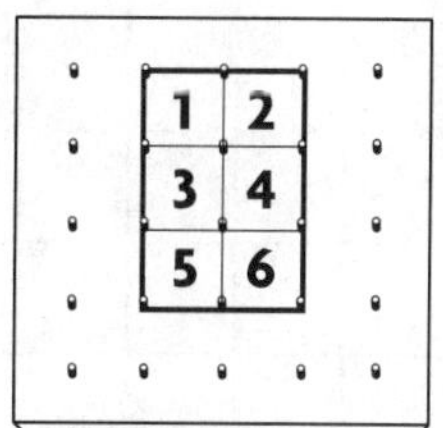

The area is 6 square units.

$\frac{1}{2}$ unit

1

$\frac{1}{2}$ unit

The area is 2 square units.

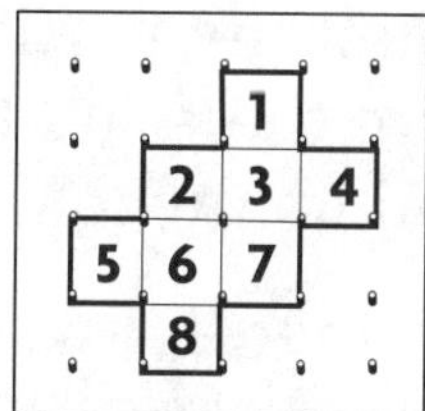

The area is 8 square units.

Find the area.

1. 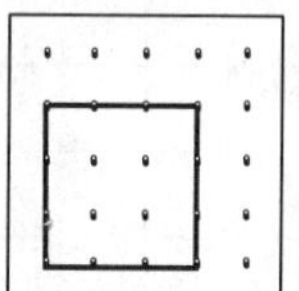
9 sq units

2. 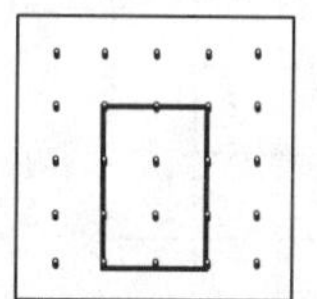
6 sq units

3. 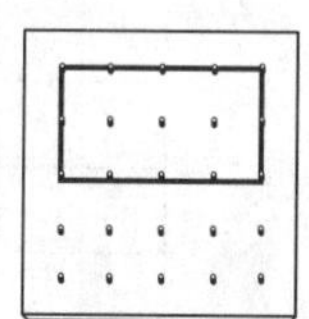
8 sq units

4. 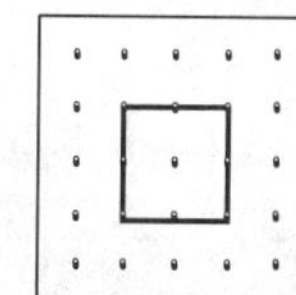
4 sq units

5. 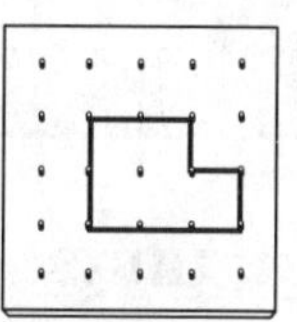
5 sq units

6. 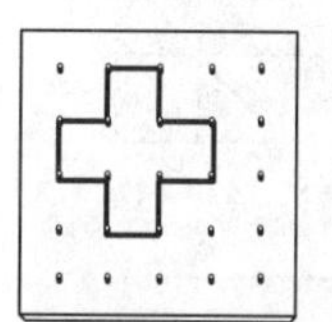
5 sq units

7. 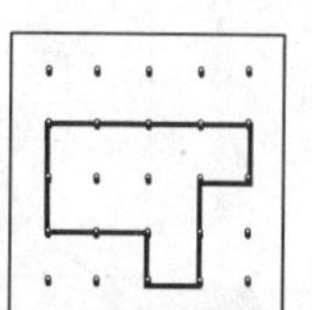
8 sq units

8. 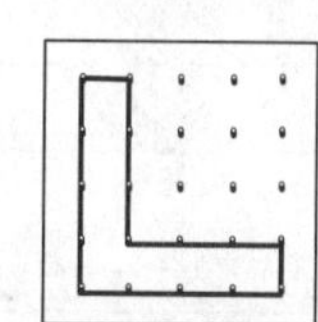
7 sq units

9. 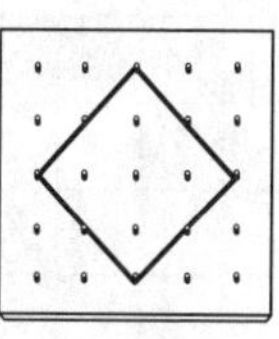
8 sq units

10. 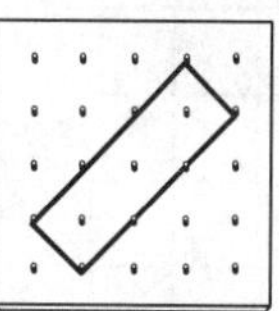
6 sq units

11. 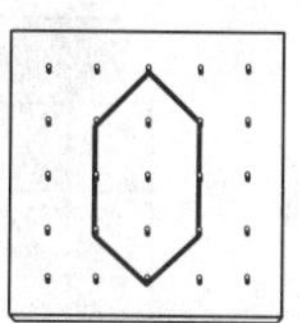
6 sq units

12. 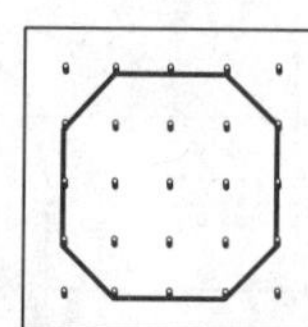
14 sq units

13. Which figures have the same area? a, b, c and d

a. 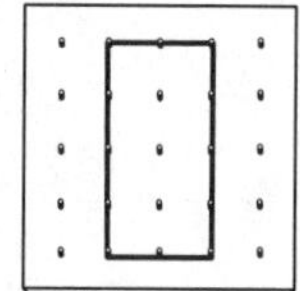

b. 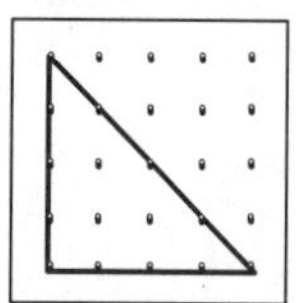

c. 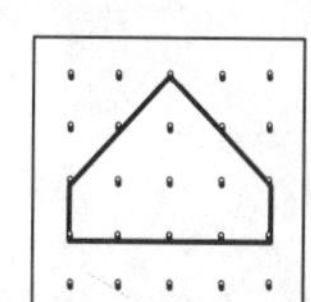

d. 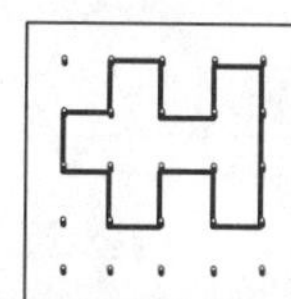

Name ____________________

Area of Irregular Figures

Here is one way to find the area of the figure.

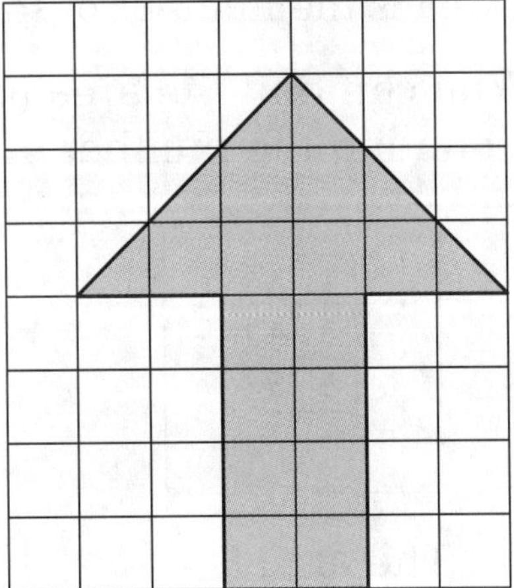

Step 1 Start by counting the whole square units. There are 14 whole square units.

Step 2 Then count the partial square units. In this figure there are 6 half units, which are equal to 3 whole units.

Step 3 Add together the whole and partial square units. $14 + 3 = 17$

So, the figure's area is 17 square units.

Find the area.

1. 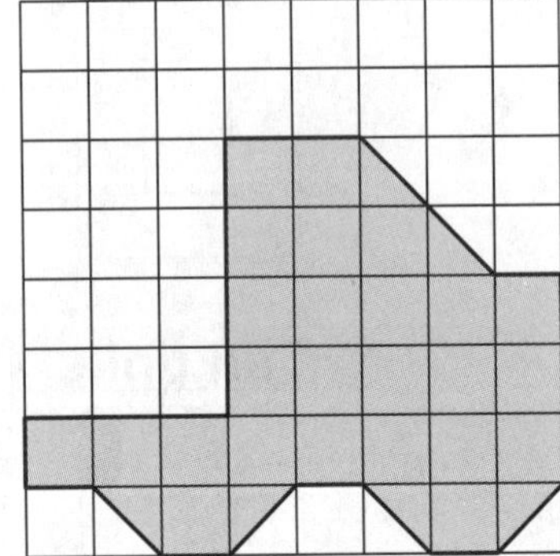

   28 sq units

2. 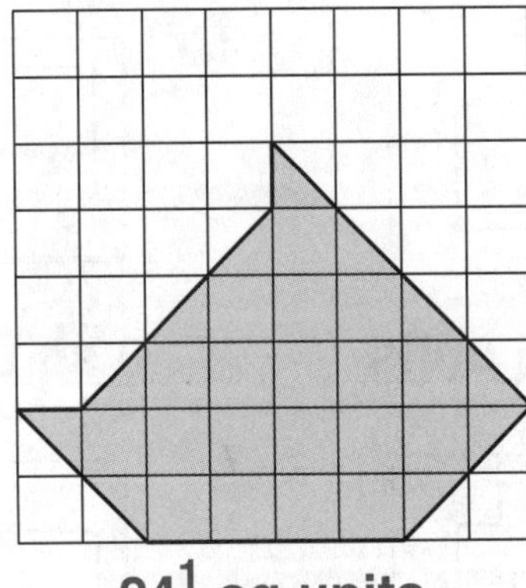

   $24\frac{1}{2}$ sq units

3. 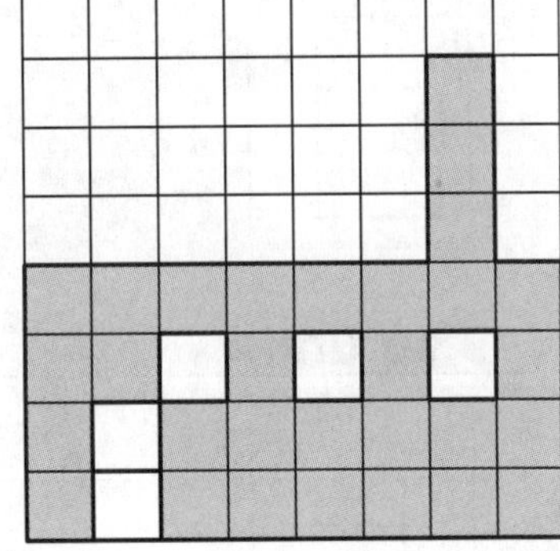

   30 sq units

4. 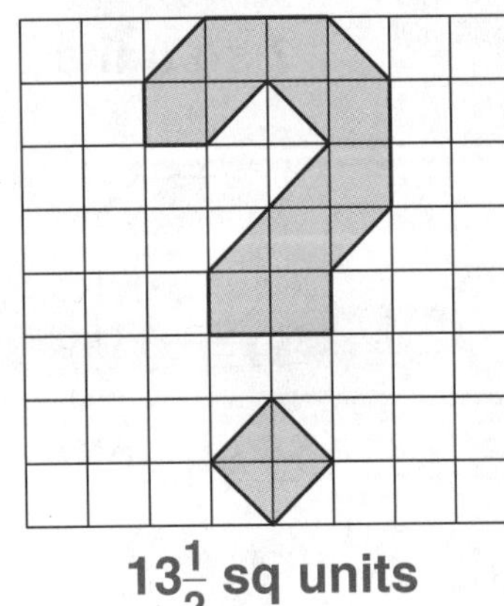

   $13\frac{1}{2}$ sq units

5. 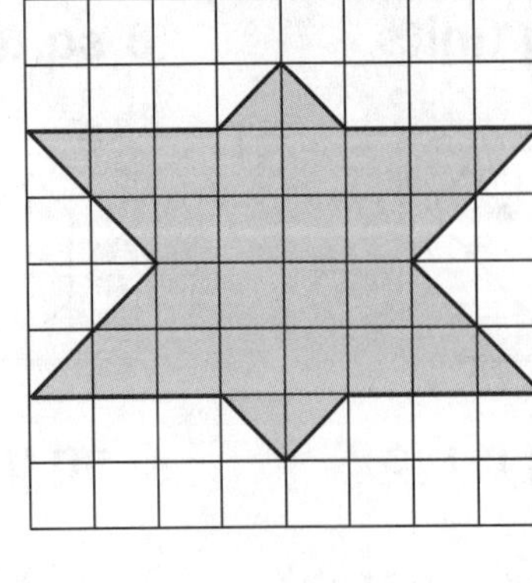

   26 sq units

6. 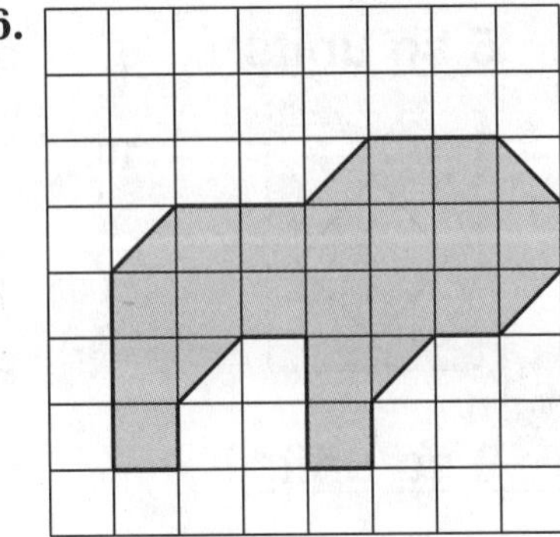

   21 sq units

Name ______________________________

Finding Area

You can find the area of a rectangle by using multiplication.

You multiply the length times the width of a rectangle to find the area.

length	×	width	=	area
7 ft	×	3 ft	=	21 sq ft

7 ft

3 ft

Write your answer in square units, such as *sq ft* or *sq yd*.

Find the area.

1. 6 in.

3 in.

18 sq in.

2. 4 yd

4 yd

16 sq yd

3. 3 km

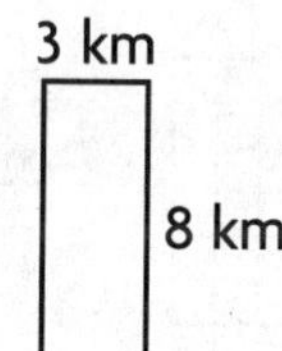

8 km

24 sq km

4. 7 ft

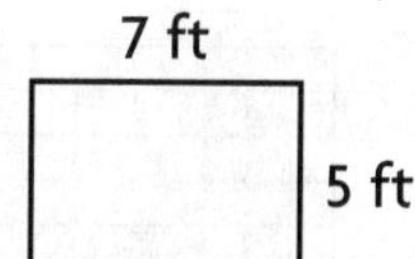

5 ft

35 sq ft

5. 9 cm

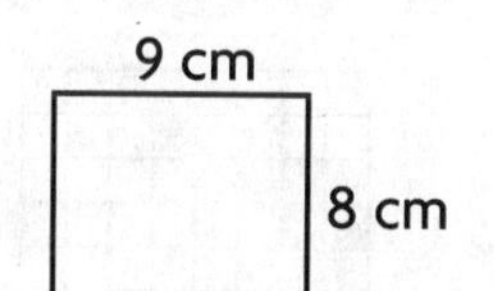

8 cm

72 sq cm

6. 8 ft

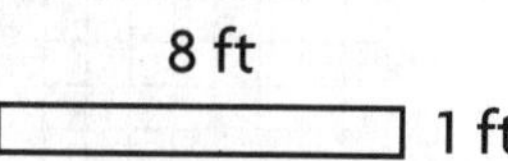

1 ft

8 sq ft

Circle the letter of the rectangle that has the greater area.

7. (a.) 6 ft, 7 ft

b. 7 ft

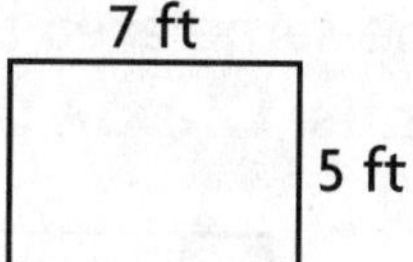

5 ft

8. a. 9 in.

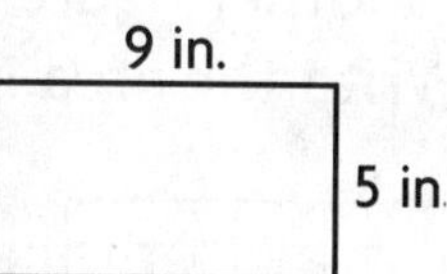

5 in.

(b.) 8 in., 7 in.

9. a. 5 km, 5 km

(b.) 9 km, 3 km

10. (a.) 6 cm, 3 cm

b. 8 cm, 2 cm

Name ______________________________

Relating Area and Perimeter

Perimeter is the distance around a figure. Perimeter is measured in linear units.

Area is the number of square units needed to cover a flat surface. Area is measured in square units.

Figures with the same area can have different perimeters.

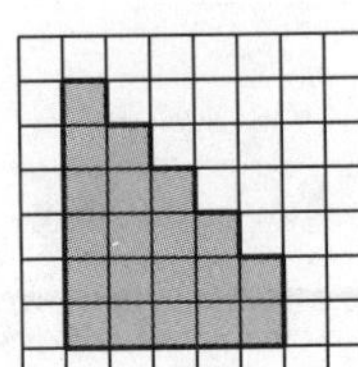

Area = 20 sq units
Perimeter = 22 units

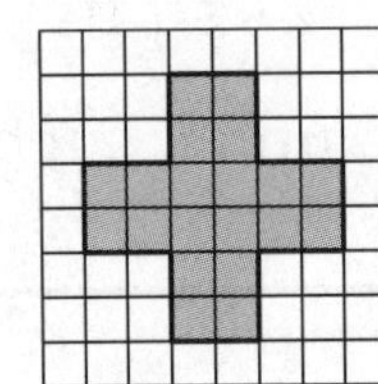

Area = 20 sq units
Perimeter = 24 units

Figures with the same perimeter can have different areas.

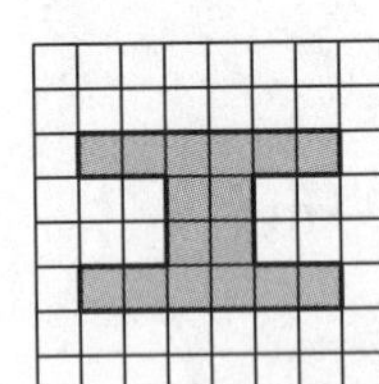

Area = 16 sq units
Perimeter = 28 units

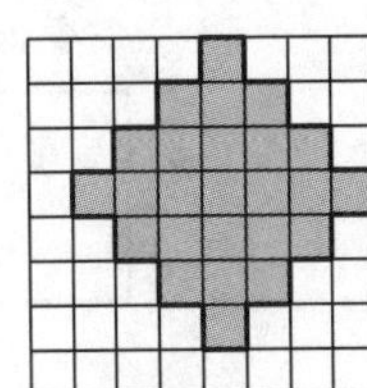

Area = 25 sq units
Perimeter = 28 units

Write the area and the perimeter.

1.

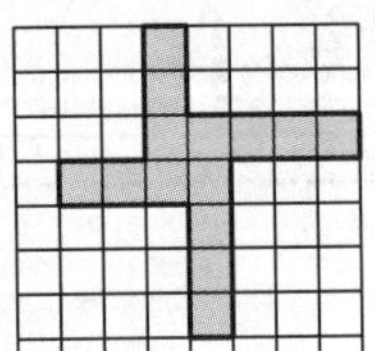

area = 14 sq units

perimeter = 28 units

2.

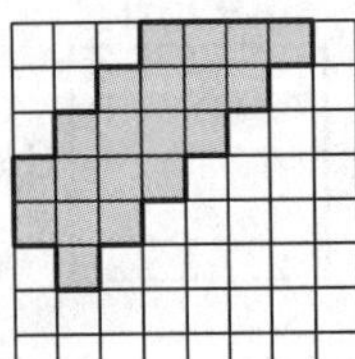

area = 20 sq units

perimeter = 26 units

3.

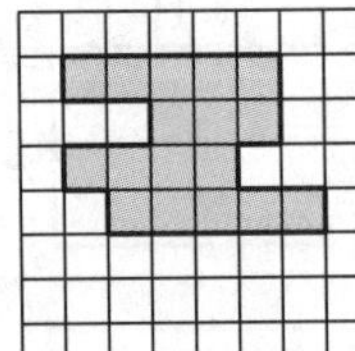

area = 17 sq units

perimeter = 26 units

For each figure, draw another figure that has the same area but a different perimeter. Write the area in square units. **Check students' drawings.**

4.

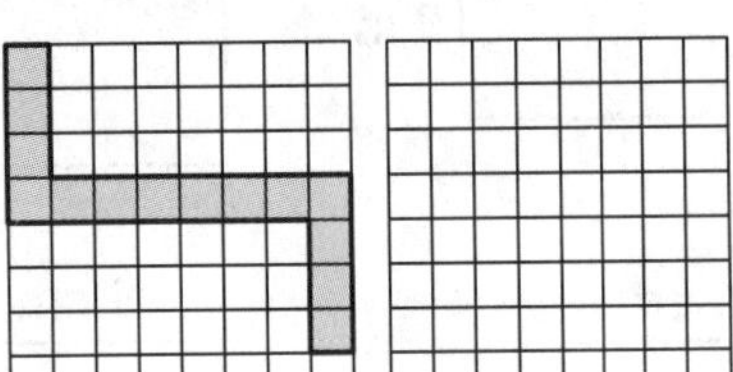

area = 14 sq units

5.

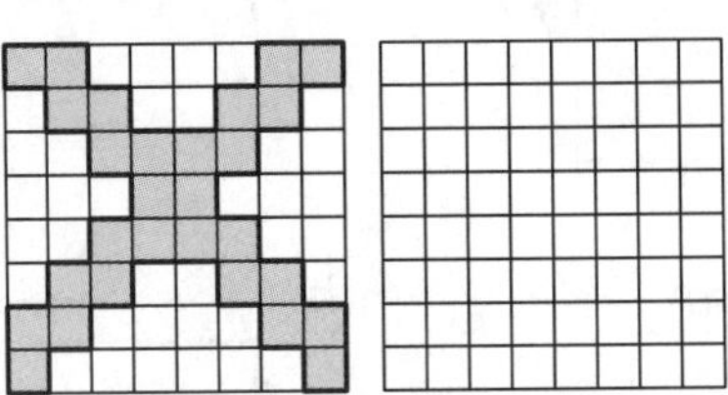

area = 28 sq units

Name ____________________

LESSON 14.5

Problem-Solving Strategy

Draw a Diagram

You can draw a diagram to solve problems involving area and perimeter. If you know the perimeter, you can figure out the greatest area.

Jane has 28 feet of fencing to go around a rectangular vegetable garden. She wants to know the shape of the rectangle that will give her the greatest area. She can draw a diagram of each shape that has a perimeter of 28 feet.

perimeter = 28 ft
area = 13 sq ft

13
1

perimeter = 28 ft
area = 24 sq ft

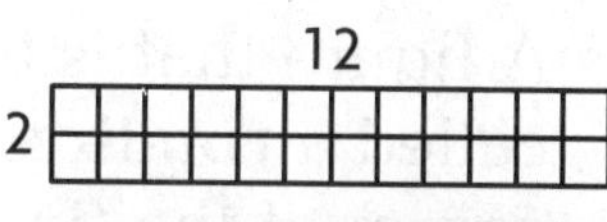

perimeter = 28 ft
area = 33 sq ft

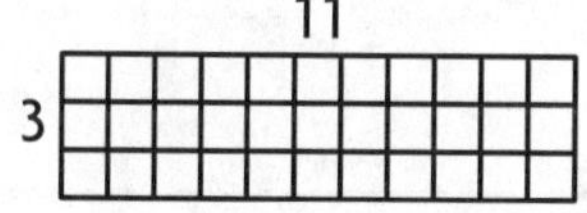

perimeter = 28 ft
area = 40 sq ft

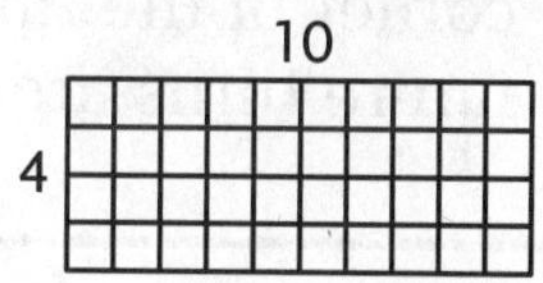

perimeter = 28 ft
area = 45 sq ft

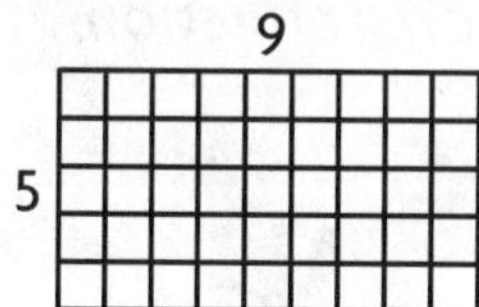

perimeter = 28 ft
area = 49 sq ft

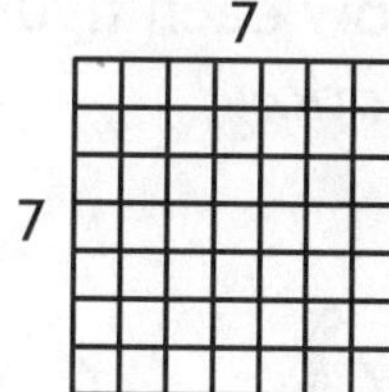

So, a square 7 feet by 7 feet has the greatest area.

Draw a diagram to solve.
Check students' diagrams.

1. Tad has 16 feet of fencing to put around his garden. What is the greatest area he can make his garden?

   16 sq ft

2. George has 36 feet of fencing to build a pen for his duck. What shape will give him the greatest area? What are the lengths of the sides?

   a square; 9 ft × 9 ft

Use the space below to draw a diagram to solve Problems 1–2.

Name ____________________

Translations, Reflections, and Rotations

When you have moved a figure, you have completed a **transformation**. Three kinds of transformations are described below.

- Sliding a figure in any direction is called a **translation.** The figure looks exactly the same but is in a new position.

- You can flip a figure over an imaginary line. The flipped figure is a mirror image called a **reflection.**

- A figure that is turned around a point is called a **rotation.** The point holds the corner of the figure while the rest of the figure turns around the point.

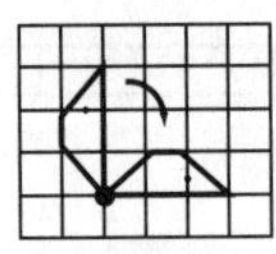
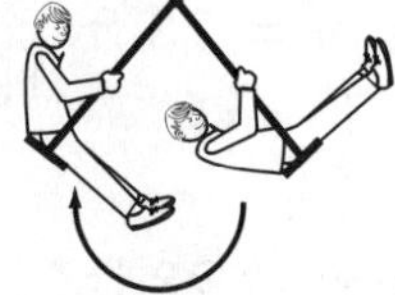

Tell how each figure was moved. Write *translation, reflection,* or *rotation.*

1. 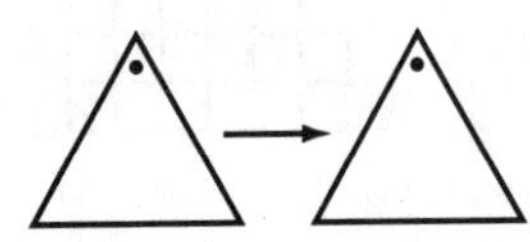
translation

2. 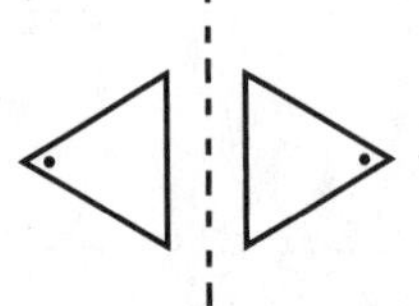
reflection

3. 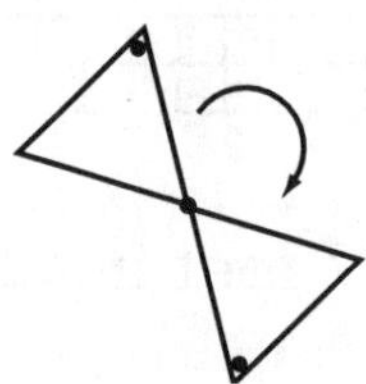
rotation

4. 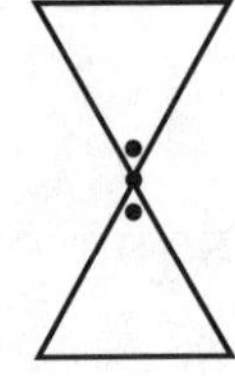
reflection or rotation

Copy each original figure. Draw figures to show a *translation, reflection,* and *rotation.* Check students' drawings. Possible answers are given.

5. Original   Translation   Reflection   Rotation

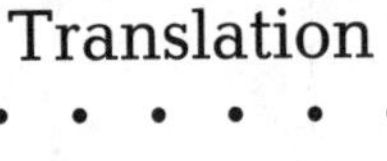
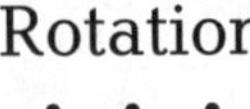
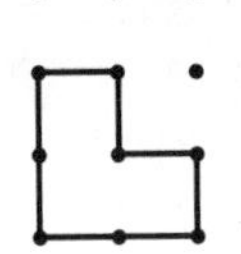
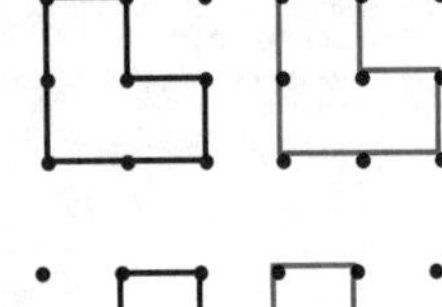
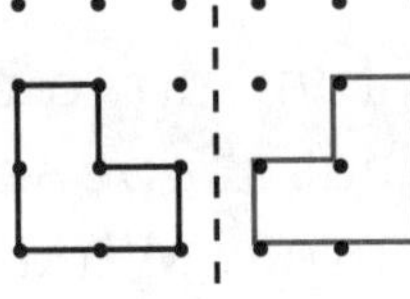
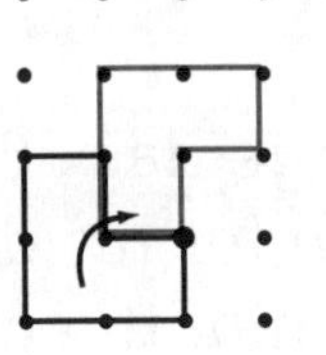

6.

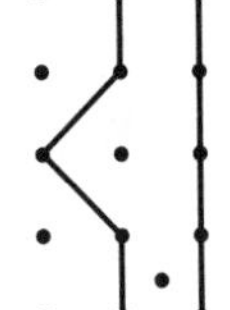
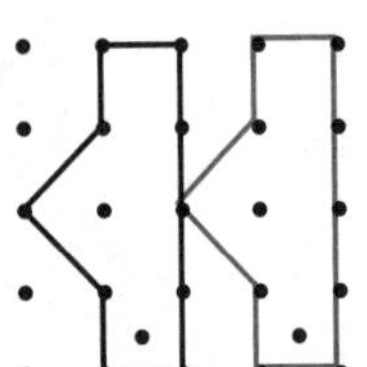
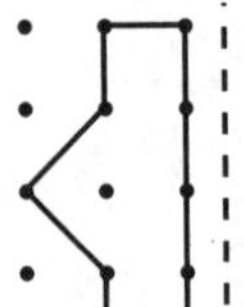
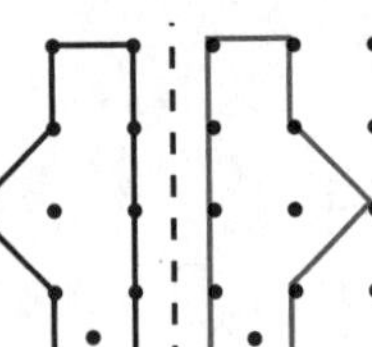
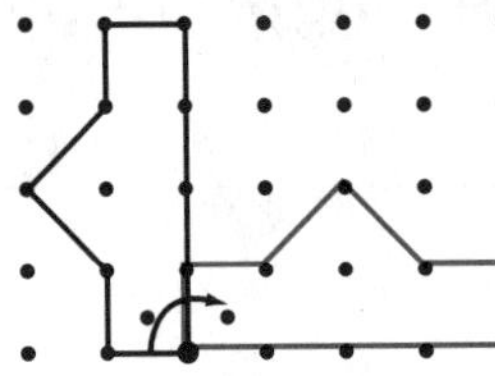

Name ____________________

LESSON 15.2

Congruence

Congruent figures have the same size and shape.

The figures at the right are congruent. They are not in the same position, but they are the same size and shape.

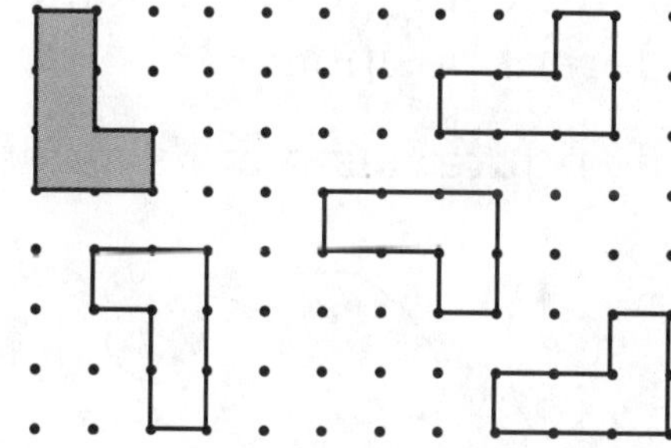

You can check for congruency by tracing one figure on another sheet of paper. Place the traced figure over the other figures and compare for congruency.

Trace one figure from each set. Place it over the second figure and check for congruency. Write whether each pair is *congruent* or *not congruent*.

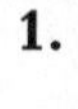

1.
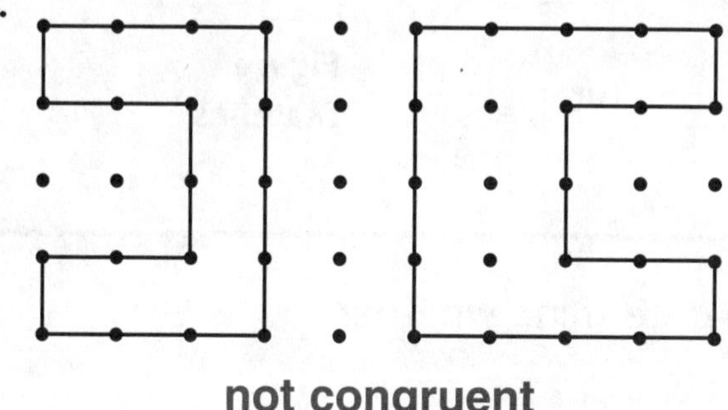

not congruent

2.
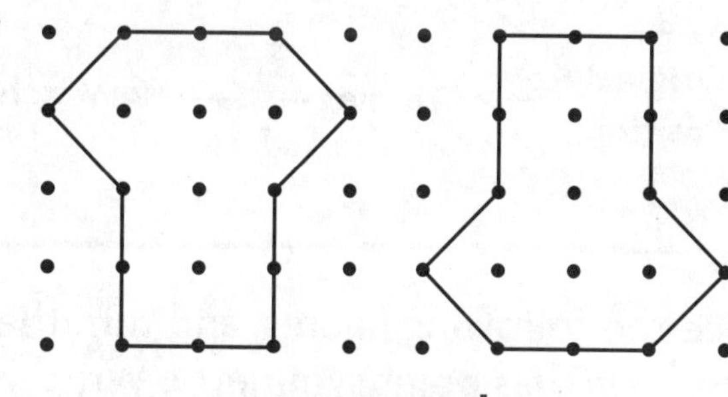

congruent

3.
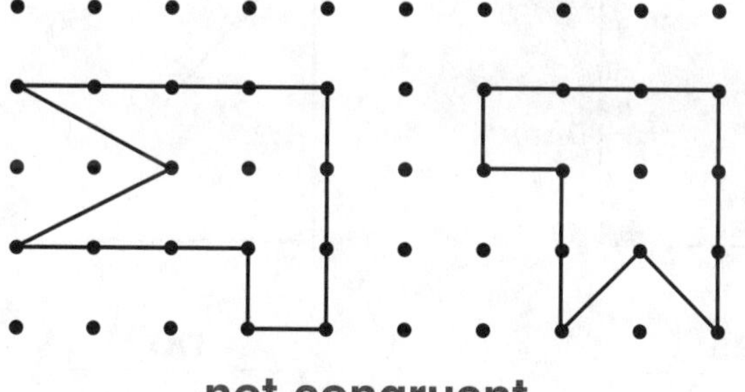

not congruent

4.
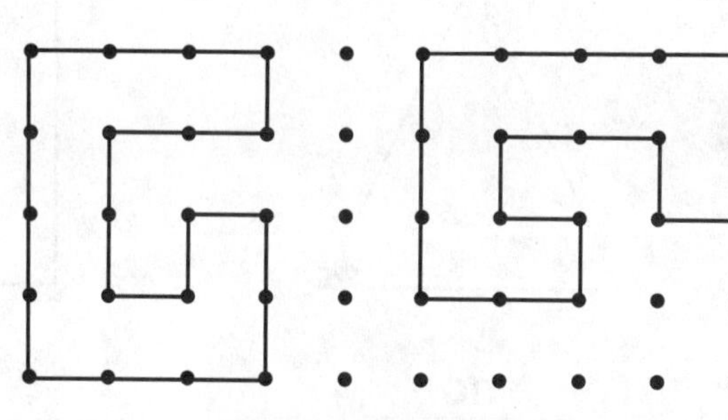

not congruent

5.
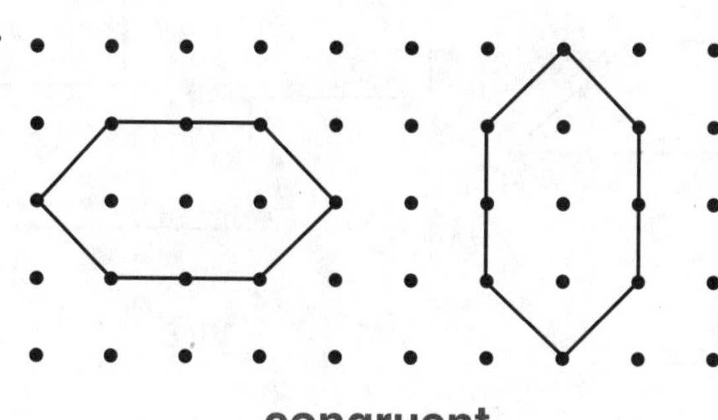

congruent

6.
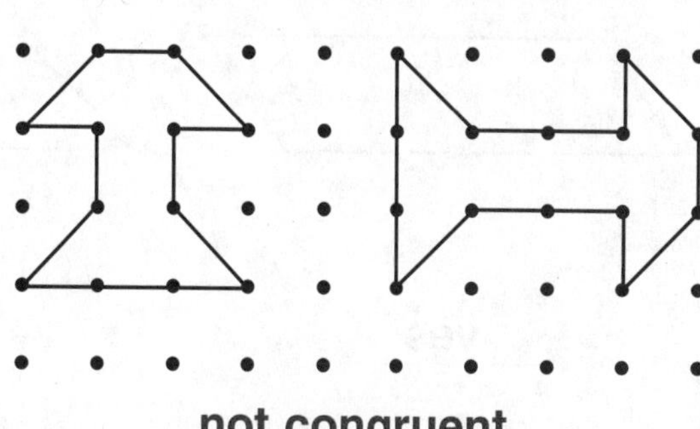

not congruent

Name ____

Two Kinds of Symmetry

A figure has **point symmetry** if it can be turned about a central point, and still look the same in at least two different positions.

This figure has point symmetry.

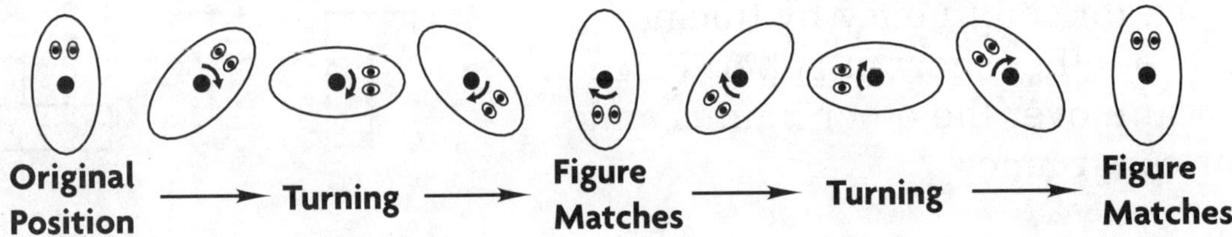

This figure *does not* have point symmetry.

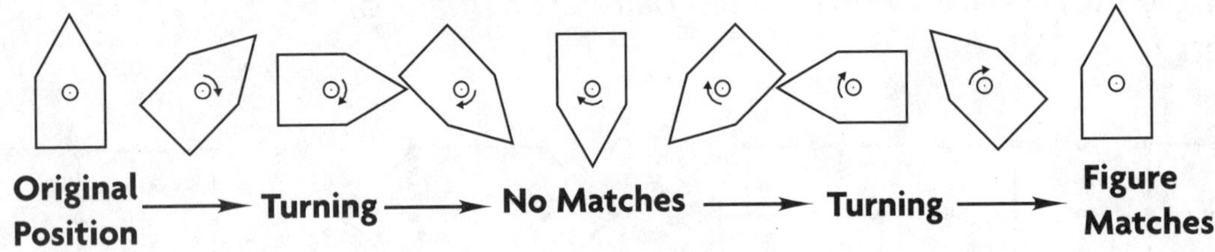

Trace the following figures, and turn them to determine whether each figure has point symmetry. Write *yes* or *no*.

1. 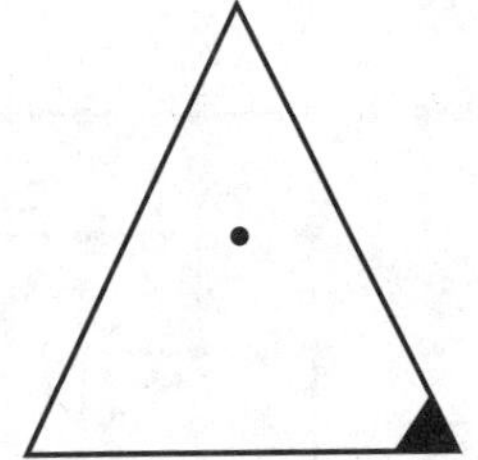

no

2. 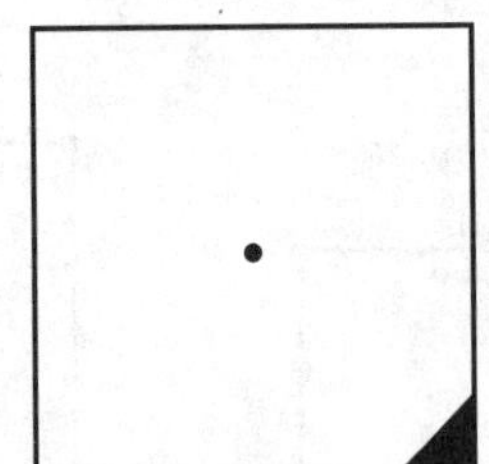

yes

3. 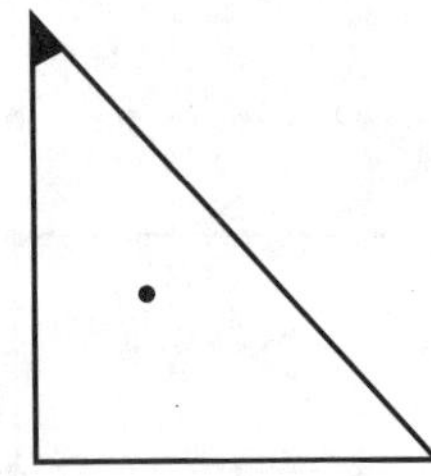

no

4. 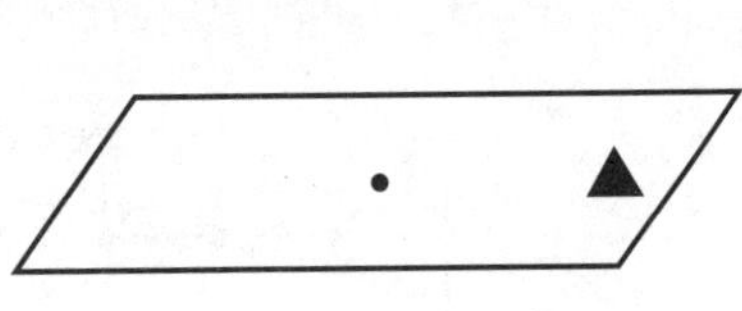

yes

5. 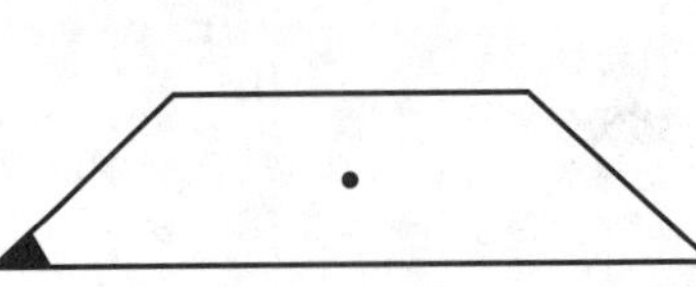

no

6. 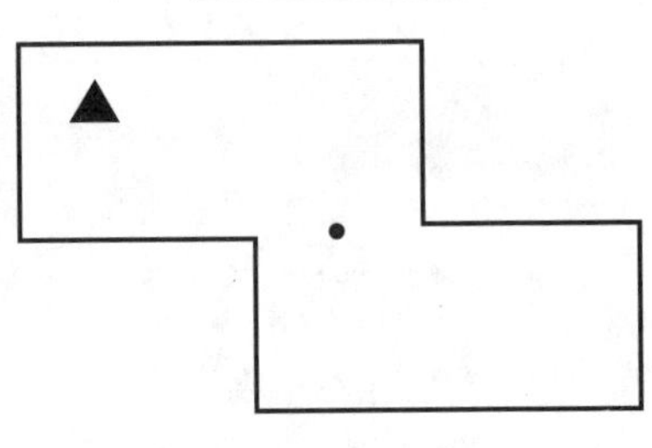

yes

Name ___________________________________

More About Symmetry

Trace and cut out the figure on the right. Fold it along the dotted line.

When you unfold the figure, you can see that one side is a reflection of the other. When a figure can be folded about a line so that its two parts are identical, the figure has **line symmetry**.

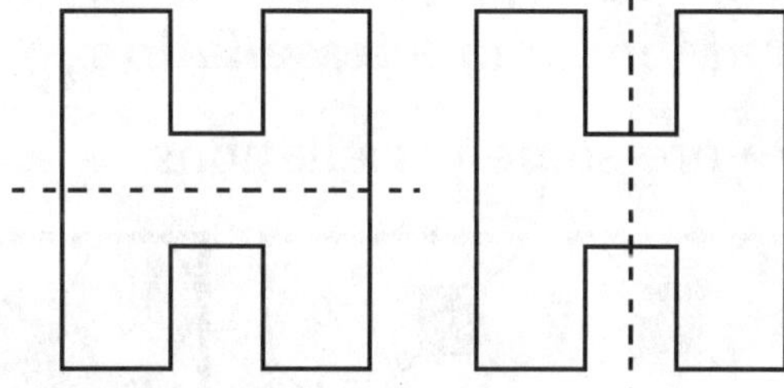

Is the dotted line a line of symmetry? Trace and cut out each figure. Fold to check. Write *yes* or *no.*

1. 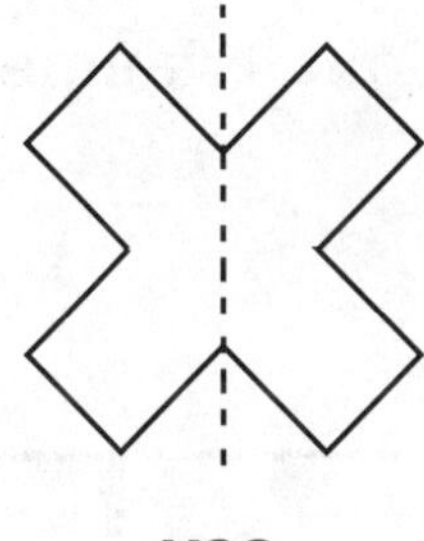
   yes

2. 
   no

3. 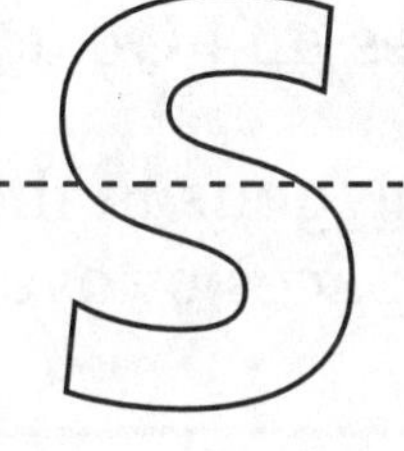

   no

4. 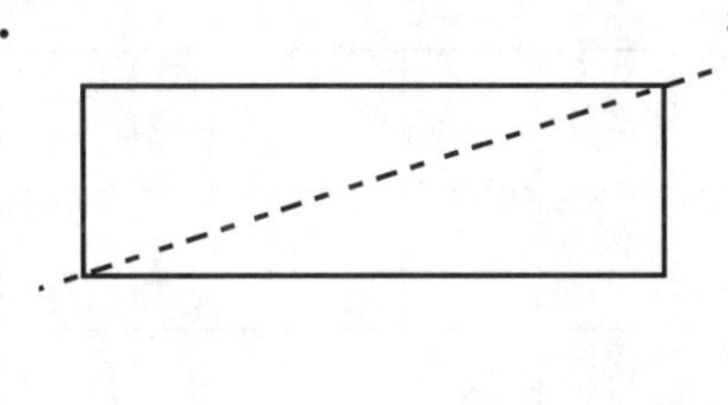
   no

5. 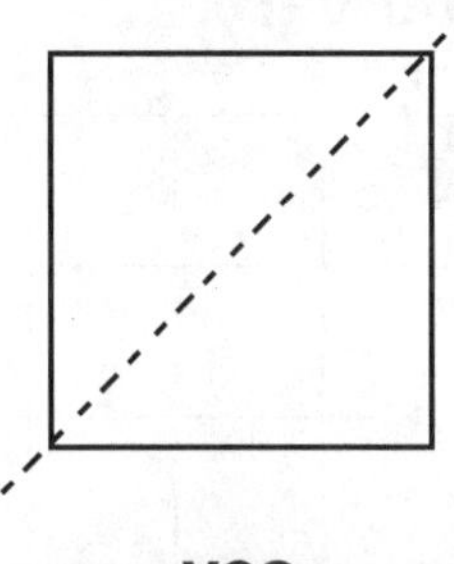
   yes

6. 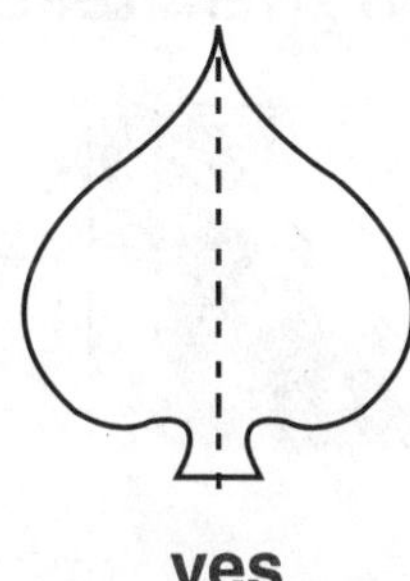
   yes

Draw the other half of the figure to show that the figure has line symmetry. **Check students' drawings.**

7. 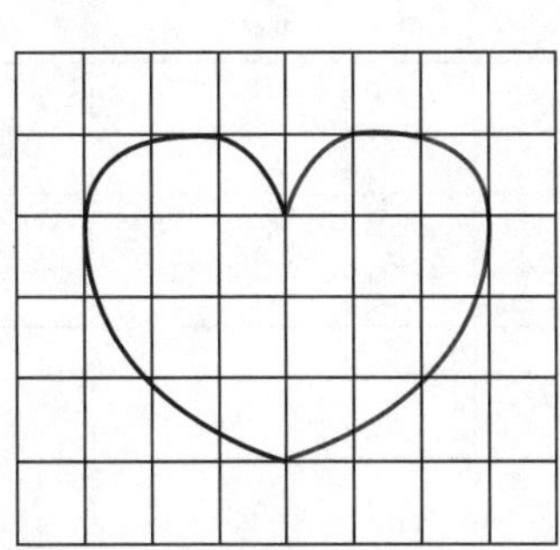

8. 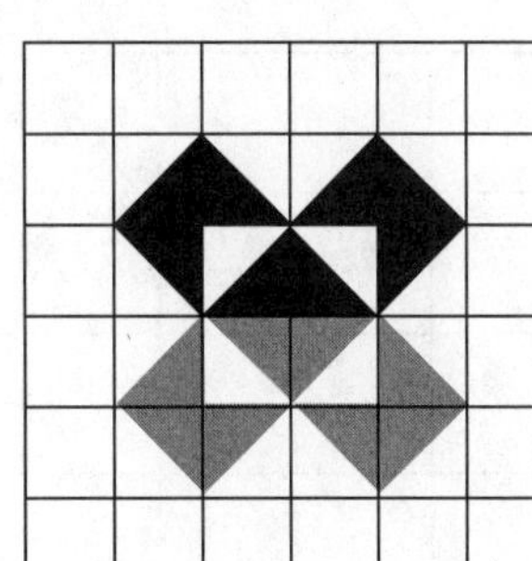

9. 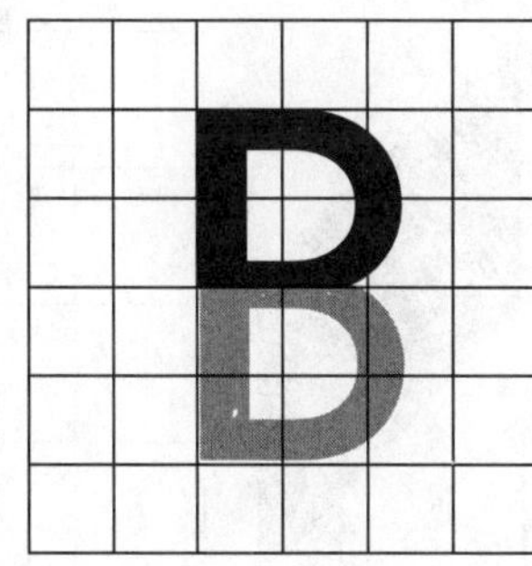

Name ______________________________

Tessellations

When you arrange polygons to cover a surface without leaving any space between them or making them overlap, you are making a **tessellation**.

Here are some tessellations:

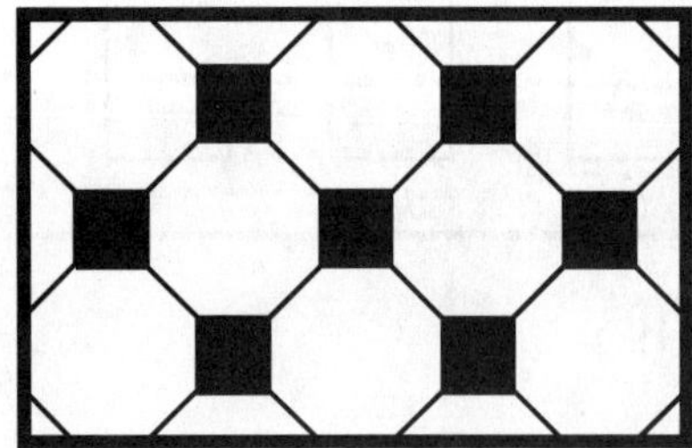

Squares and Octagons

Hexagons

Right Triangles

The polygons in these tessellations all connect with no open spaces or overlaps.

Use two different colors to make your own tessellation. Use the polygons given. **Tessellations will vary.**

1.

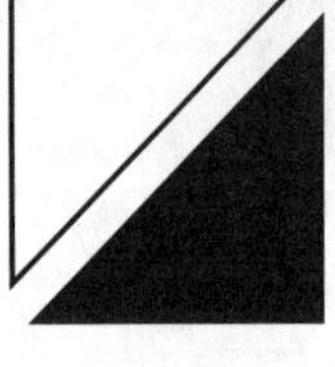

2.

Name ____________________

Changing Sizes of Shapes

Similar figures do not have to be the same size, but they have to be the same shape. To change the size of a figure, follow these steps.

Step 1
Place corner points on the figure.
Count the spaces between each point.

Step 2
Count the same number of spaces on a smaller or a larger grid. Place the points.

Step 3
Complete the sides of the similar figure.

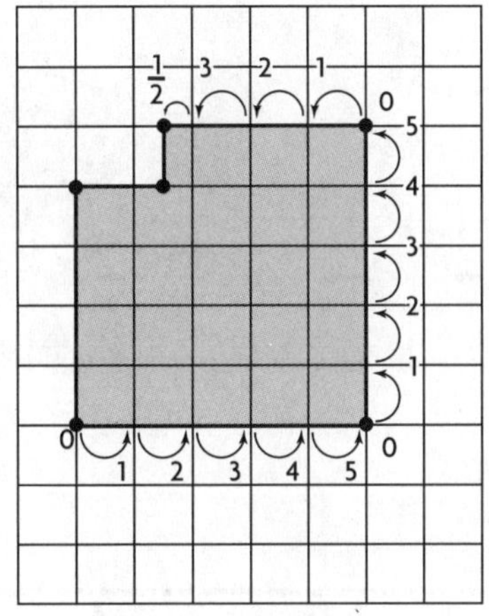

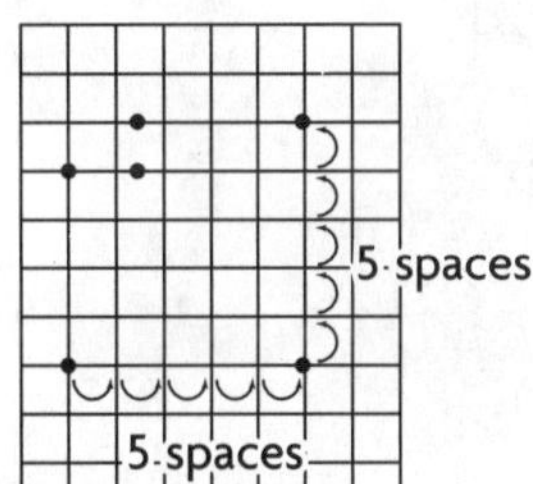

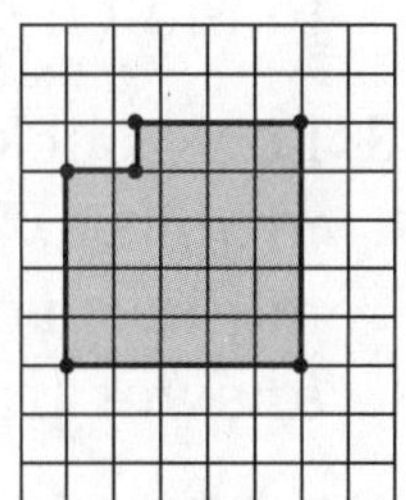

Reduce each figure by using the grid below.

1.

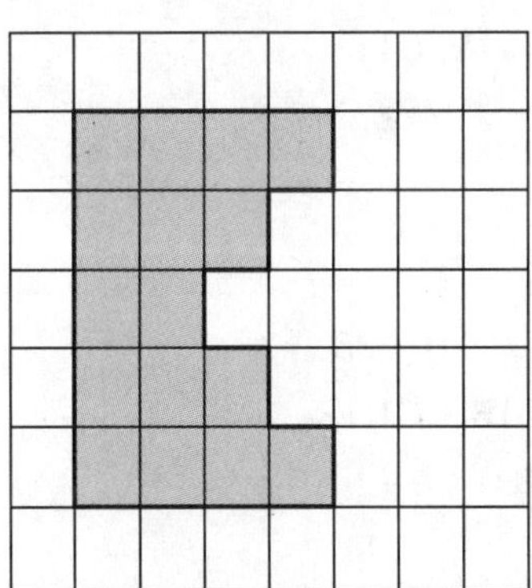

2.

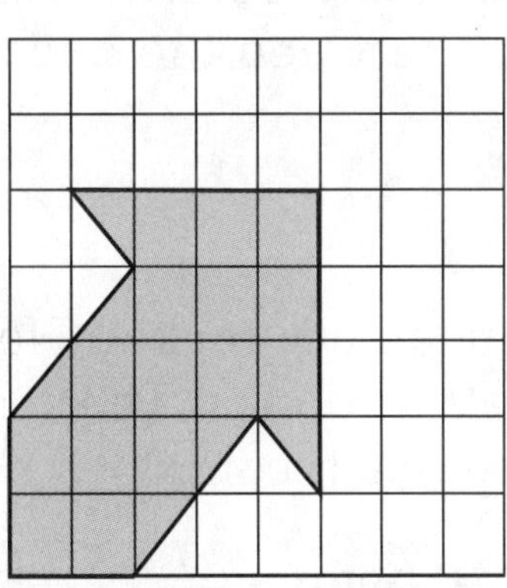

3.

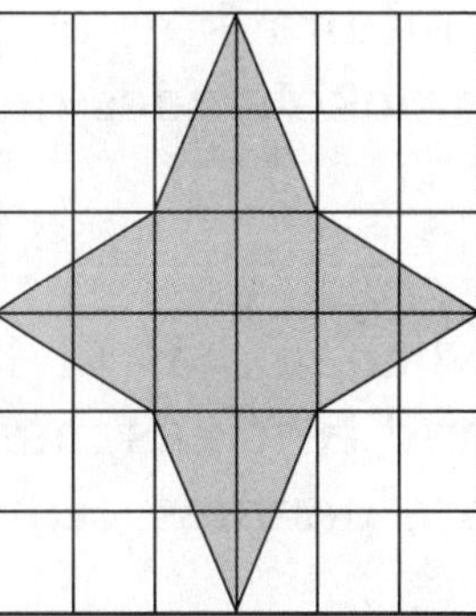

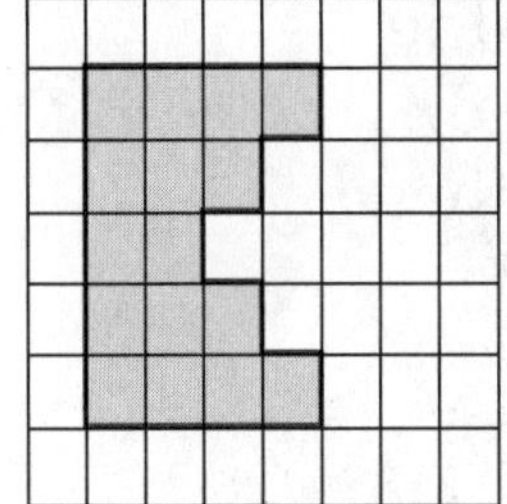

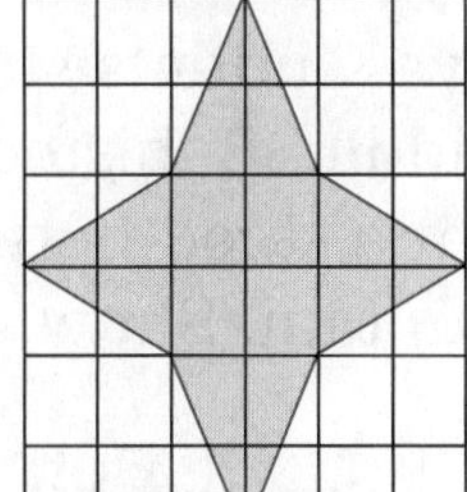

Name ______________________________

Problem-Solving Strategy

Make a Model

When you need to solve a problem involving geometric figures, it may be helpful to make a model.

Fred is building a fence 10 yards long. He will place a post at each end of the fence and posts every 2 yards between the ends. How many fence posts does he need?

Step 1 Draw a straight line to represent the fence.

Step 2 Use models to represent the fence posts. Move the models around to match the problem.

Step 3 Look back at the problem. If your model matches, answer the question.

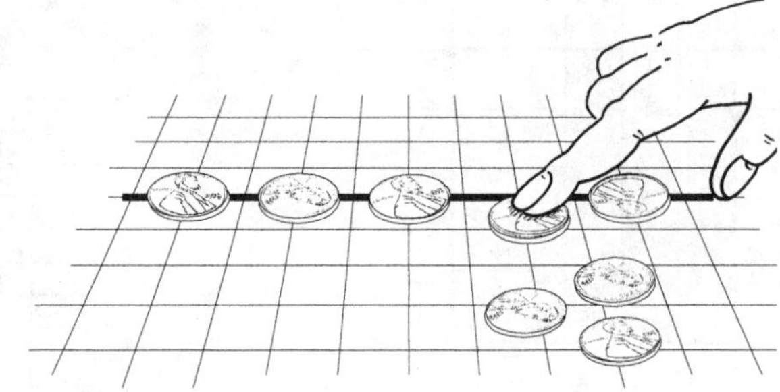

Fred needs 6 posts.

Make a model to solve.

1. T.J. is putting up a 12-foot shelf in the garage. He will hang the shelf with a bracket at each end of the shelf and brackets every 3 feet between the ends. How many brackets does he need?

   5 brackets

2. Michael has $1.10 in coins in his pocket. He has 5 dimes and twice as many nickels as dimes. The rest of his coins are pennies. How many of each coin does he have?

   5 dimes, 10 nickels, 10 pennies.

3. Five students stood in a line for water. Holly was in the middle. Bob stood next to Holly. Chris was at the end. Anne stood between Holly and Chris. Mark was at the fountain. Show the students' order.

   Mark, Bob, Holly, Anne, Chris

Name ______________________

LESSON 16.1

Multiplying by Multiples

A **multiple** is the product of a given number and another whole number. For example, 60 is a multiple of 6 and 10: $6 \times 10 = 60$. You can use basic multiplication facts and mental math to find the product.

Solve $60 \times 2 = \underline{\ ?\ }$.

Step 1 Think of the basic fact. Multiply. $6 \times 2 = 12$

Step 2 Count the number of zeros in the factors. 6<u>0</u> has 1 zero.

Step 3 Add the number of zeros in the factors to the basic-fact product. The answer is 12<u>0</u>.

Write the basic multiplication fact.

1. $70 \times 4 = 280$ — 7 × 4 = 28
2. $2{,}000 \times 2 = 4{,}000$ — 2 × 2 = 4
3. $600 \times 8 = 4{,}800$ — 6 × 8 = 48

Finish each pattern.

4. $3 \times 3 =$ 9
   $30 \times 3 =$ 90
   $300 \times 3 =$ 900
   $3{,}000 \times 3 =$ 9,000

5. $7 \times 7 =$ 49
   $70 \times 7 =$ 490
   $700 \times 7 =$ 4,900
   $7{,}000 \times 7 =$ 49,000

6. $5 \times 4 =$ 20
   $50 \times 4 =$ 200
   $500 \times 4 =$ 2,000
   $5{,}000 \times 4 =$ 20,000

7. $6 \times 2 =$ 12
   $60 \times 2 =$ 120
   $600 \times 2 =$ 1,200
   $6{,}000 \times 2 =$ 12,000

8. $4 \times 8 =$ 32
   $40 \times 8 =$ 320
   $400 \times 8 =$ 3,200
   $4{,}000 \times 8 =$ 32,000

9. $3 \times 9 =$ 27
   $30 \times 9 =$ 270
   $300 \times 9 =$ 2,700
   $3{,}000 \times 9 =$ 27,000

10. $3 \times$ 6 $= 18$
    30 $\times 6 = 180$
    300 $\times 6 = 1{,}800$
    $3{,}000 \times 6 =$ 18,000

11. $7 \times 5 =$ 35
    70 $\times 5 = 350$
    $700 \times 5 =$ 3,500
    $7{,}000 \times$ 5 $= 35{,}000$

12. $9 \times$ 4 $= 36$
    90 $\times 4 = 360$
    $900 \times 4 =$ 3,600
    9,000 $\times 4 = 36{,}000$

Name ______________________

LESSON 16.2

A Way to Multiply

Suppose you have 3 boxes of golf balls. Each box contains 25 balls. How many golf balls do you have?

You can solve the problem by using base-ten blocks.

After regrouping the ones, you have 7 tens and 5 ones, or 75. So, you have 75 golf balls.

You can also solve the problem by using a repeated addition number sentence. You know you have 3 groups of 25.
So, $25 + 25 + 25 = 75$, or $3 \times 25 = 75$.

Write a repeated addition number sentence for each problem. Then write a multiplication number sentence. **Order of number sentences may vary.**

1. Paul's Pizzeria ordered 7 boxes of cheese. Each box contained 50 pounds of cheese. How much did the 7 boxes of cheese weigh?
   50 + 50 + 50 + 50 + 50 + 50 + 50 = 350;
   7 × 50 = 350; 350 lb

2. In Mr. Palmer's class, 6 students each read 26 books for a reading contest. How many books in all did the 6 students read?
   26 + 26 + 26 + 26 + 26 + 26 = 156;
   6 × 26 = 156; 156 books

3. The 10 members of the school's track club ran a 6-mile race. How many miles in all did the 10 members run?
   6 + 6 + 6 + 6 + 6 + 6 + 6 + 6 + 6 + 6 = 60;
   10 × 6 = 60; 60 mi

4. A farm donated 9 boxes of fruit to a local food bank. Each box contained 48 pieces of fruit. How many pieces of fruit did the farm donate in all?
   48 + 48 + 48 + 48 + 48 + 48 + 48 + 48 + 48 = 432;
   9 × 48 = 432; 432 pieces of fruit

Find the product.

5. $37 \times 6 =$ **222**

6. $69 \times 4 =$ **276**

7. $42 \times 4 =$ **168**

8. $27 \times 8 =$ **216**

Name ______________________________

Modeling Multiplication

Solve. $17 \times 6 =$?, or $\begin{array}{r} 17 \\ \times\ 6 \\ \hline \end{array}$

- You can find the product by breaking the two-digit number apart.
- Multiply, using the simpler numbers.

First, multiply 6 times 7. $6 \times 7 = 42$ — partial products

Second, multiply 6 times 10. $6 \times 10 = 60$ — partial products

Third, add the **partial products** to find the product.
$42 + 60 = 102$

Find each product by using partial products. Show the three steps you used.

1. $\begin{array}{r} 41 \\ \times\ 6 \\ \hline \end{array}$

   $6 \times 1 = 6$

   $6 \times 40 = 240$

   $6 + 240 = 246$

2. $\begin{array}{r} 37 \\ \times\ 3 \\ \hline \end{array}$

   3 × 7 = 21

   3 × 30 = 90

   21 + 90 = 111

3. $\begin{array}{r} 67 \\ \times\ 2 \\ \hline \end{array}$

   2 × 7 = 14

   2 × 60 = 120

   14 + 120 = 134

4. $\begin{array}{r} 55 \\ \times\ 4 \\ \hline \end{array}$

   4 × 5 = 20

   4 × 50 = 200

   20 + 200 = 220

5. $\begin{array}{r} 29 \\ \times\ 7 \\ \hline \end{array}$

   7 × 9 = 63

   7 × 20 = 140

   63 + 140 = 203

6. $\begin{array}{r} 32 \\ \times\ 5 \\ \hline \end{array}$

   5 × 2 = 10

   5 × 30 = 150

   10 + 150 = 160

7. $\begin{array}{r} 46 \\ \times\ 6 \\ \hline \end{array}$

   6 × 6 = 36

   6 × 40 = 240

   36 + 240 = 276

8. $\begin{array}{r} 71 \\ \times\ 3 \\ \hline \end{array}$

   3 × 1 = 3

   3 × 70 = 210

   3 + 210 = 213

9. $\begin{array}{r} 54 \\ \times\ 8 \\ \hline \end{array}$

   8 × 4 = 32

   8 × 50 = 400

   32 + 400 = 432

Name ______________________________

Recording Multiplication

Find the product 2 × 434.

Step 1

Multiply the ones.

H	T	O
4	3	**4**
×		**2**
		8

Step 2

Multiply the tens.

H	T	O
4	**3**	4
×		**2**
	6	8

Step 3

Multiply the hundreds.

H	T	O
4	3	4
×		**2**
8	6	8

So, 2 × 434 = 868.

Find the product 4 × 367. You may need to regroup.

Step 1

Multiply the ones.
Regroup?

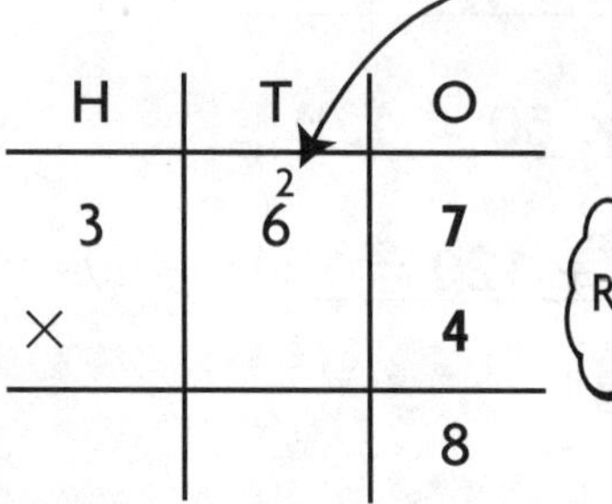

H	T	O
3	$\overset{2}{6}$	**7**
×		**4**
		8

Step 2

Multiply the tens.
Add the regrouped
tens. Regroup?

H	T	O
$\overset{2}{3}$	$\overset{2}{\mathbf{6}}$	7
×		**4**
	6	8

Regroup
26

Step 3

Multiply the hundreds.
Add the regrouped
hundreds.

H	T	O
$\overset{2}{\mathbf{3}}$	$\overset{2}{6}$	7
×		**4**
14	6	8

So, 4 × 367 = 1,468.

Complete.

1.

H	T	O
1	2	3
×		3
3	6	9

2.

H	T	O
3	1	1
×		4
12	4	4

3.

H	T	O
$\overset{1}{4}$	$\overset{4}{2}$	7
×		6
25	6	2

Name ______________________

Practicing Multiplication

Use estimation when you multiply with problems involving money.

Find the product 4 × $2.99.

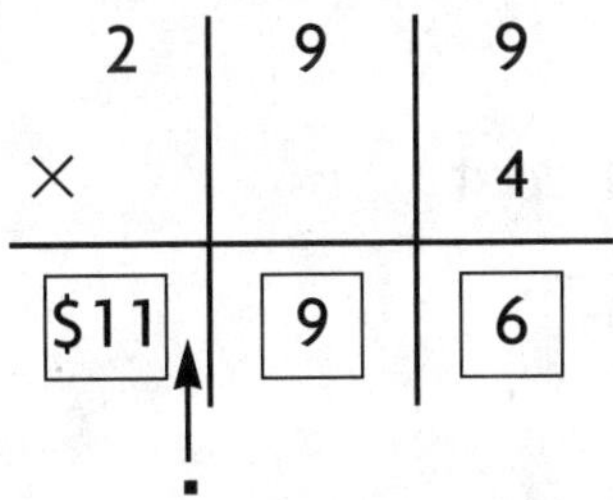

- First, estimate your answer. 4 × $3.00 is $12.00.
- Second, multiply by using whole numbers.
- Third, compare your answer with your estimate. Use your estimate to help you place the decimal point and dollar sign.
- Since 1196 is close to 1200, you place the decimal point between the 1 and the 9. Add the dollar sign.

Estimate the product. Then place the dollar sign and decimal point in each given product. **Estimates may vary.**

1. $6.62 × 6 = $39.72 — Estimate: **$42.00**

2. $9.34 × 3 = $28.02 — Estimate: **$27.00**

3. $7.54 × 7 = $52.78 — Estimate: **$56.00**

Multiply and record each product.

4. $1.74 × 5 = **$8.70**

5. $2.23 × 7 = **$15.61**

6. $4.92 × 6 = **$29.52**

7. $3.37 × 4 = **$13.48**

8. $2.49 × 8 = **$19.92**

9. $3.87 × 3 = **$11.61**

10. $1.98 × 5 = **$9.90**

11. $4.21 × 6 = **$25.26**

12. $7.13 × 4 = **$28.52**

Name ______________________________

LESSON 16.5

Problem-Solving Strategy

Write a Number Sentence

Writing a number sentence helps organize the information within a math problem. Consider the following word problem. Bill wants to buy 7 sets of collectors' cards. Each set cost \$2.99. How much money does Bill need?

number of sets	×	cost per set	=	total cost
7	×	\$2.99	=	n

$7 \times \$2.99 = n$ $n = \$20.93$

Write a number sentence to solve the problem.

1. Leslie buys 5 books. The books cost \$3.99 each. How much does her purchase cost in all?

   $5 \times \$3.99 = n$ $n = \$19.95$

2. Mark exercises for 50 minutes a day, 5 days a week. How many minutes each week does Mark exercise?

   $5 \times 50 = n$ $n = 250$ min

3. Last year, Lorna's mother traveled 7,496 miles on business trips. This year, she traveled 6,957 miles. How many more miles did Lorna's mother travel last year?

   $7{,}496 - 6{,}957 = n$ $n = 539$ more miles

4. Emma paid \$39.95 for a CD, \$19.95 for a user's manual, and \$79.95 for a game stick. How much did she spend in all?

   $\$39.95 + \$19.95 + \$79.95 = n$ $n = \$139.85$

5. Joy practices the piano for 35 minutes each day. How many minutes does she practice in one week?

   $7 \times 35 = n$ $n = 245$ minutes

Name ______________________________

Patterns with Multiples

If you can multiply ones, then you can multiply tens.

$$\begin{array}{r} 2 \\ \times 3 \\ \hline 6 \end{array} \qquad \begin{array}{l} 2 \text{ tens} \\ \times 3 \\ \hline 6 \text{ tens} \end{array} = \begin{array}{r} 20 \\ \times 3 \\ \hline 60 \end{array}$$

You can also multiply hundreds.

$$\begin{array}{r} 4 \\ \times 2 \\ \hline 8 \end{array} \qquad \begin{array}{l} 4 \text{ hundreds} \\ \times 2 \\ \hline 8 \text{ hundreds} \end{array} = \begin{array}{r} 400 \\ \times 2 \\ \hline 800 \end{array}$$

You can even multiply thousands!

$$\begin{array}{r} 6 \\ \times 6 \\ \hline 36 \end{array} \qquad \begin{array}{l} 6 \text{ thousands} \\ \times 6 \\ \hline 36 \text{ thousands} \end{array} = \begin{array}{r} 6,000 \\ \times 6 \\ \hline 36,000 \end{array}$$

Complete.

	Ones	Tens		Hundreds		Thousands	
1.	6 × 4 = **24**	6 tens × 4 = **24 tens**	60 × 4 = **240**	6 hundreds × 4 = **24 hundreds**	600 × 4 = **2,400**	6 thousands × 4 = **24 thousands**	6,000 × 4 = **24,000**
2.	7 × 6 = **42**	7 tens × 6 = **42 tens**	70 × 6 = **420**	7 hundreds × 6 = **42 hundreds**	700 × 6 = **4,200**	7 thousands × 6 = **42 thousands**	7,000 × 6 = **42,000**
3.	8 × 5 = **40**	8 tens × 5 = **40 tens**	80 × 5 = **400**	8 hundreds × 5 = **40 hundreds**	800 × 5 = **4,000**	8 thousands × 5 = **40 thousands**	8,000 × 5 = **40,000**
4.	9 × 6 = **54**	9 tens × 6 = **54 tens**	90 × 6 = **540**	9 hundreds × 6 = **54 hundreds**	900 × 6 = **5,400**	9 thousands × 6 = **54 thousands**	9,000 × 6 = **54,000**
5.	4 × 8 = **32**	4 tens × 8 = **32 tens**	40 × 8 = **320**	4 hundreds × 8 = **32 hundreds**	400 × 8 = **3,200**	4 thousands × 8 = **32 thousands**	4,000 × 8 = **32,000**
6.	5 × 7 = **35**	5 tens × 7 = **35 tens**	50 × 7 = **350**	5 hundreds × 7 = **35 hundreds**	500 × 7 = **3,500**	5 thousands × 7 = **35 thousands**	5,000 × 7 = **35,000**

Name ______________________________

LESSON 17.1

Problem-Solving Strategy

Find a Pattern

Carrie has finished part of her square quilt. How many patches of each color will be in the final quilt?

B	R	W	B	R	W	B	R	W
R	W	B	R	W	B	R	W	B
B	R	W	B	R	W	B	R	W
R	W	B	R	W	B	R	W	B
B	R	W	B	R	W	B	R	W
R	W	B	R	W	B	R	W	B

B = Blue R = Red W = White

Look for patterns to help you make a plan for solving.

Since this is a square quilt and there are 9 patches along the top, you know there must be 9 rows.

Look closely at the quilt. There is a pattern of 3 blues, 3 reds, and 3 whites in each row. Multiply each color by the 9 rows. So, there are 27 blue, 27 red, and 27 white patches.

Blue	Red	White
$3 \times 9 = 27$	$3 \times 9 = 27$	$3 \times 9 = 27$

Find a pattern and solve.

1. Shade the figure to finish the pattern.

2. There are an equal number of trucks and cars in the lot. Each truck has 8 wheels, and each car has 4 wheels. How many trucks and cars are there if there are 36 wheels?

3 trucks, 3 cars

3. The Kopi Center is open from 9 A.M. to 7 P.M., 7 days a week. How many hours is it open in a 30-day month?

300 hr

Name ________________________________

LESSON 17.2

Estimating Products with Multiples

When you don't need an exact answer, you can estimate by rounding the factors.

To round:

- Circle the digit to be rounded.
- Underline the digit to its right.
- If the underlined digit is 5 or greater, increase the circled digit by 1. Otherwise, the circled digit stays the same.
- Write zeros in all places to the right of the circled digit.

(5)51 → 600 ← rounded up
× (2)4 → × 20 ← rounded down
12,000

Rewrite each problem by rounding each factor. Then estimate the product.

1. 62 → 60; × 24 → × 20; 1,200
2. 43 → 40; × 37 → × 40; 1,600
3. 75 → 80; × 24 → × 20; 1,600
4. 28 → 30; × 17 → × 20; 600
5. 44 → 40; × 23 → × 20; 800
6. 94 → 90; × 36 → × 40; 3,600
7. \$37 → \$40; × 34 → × 30; \$1,200
8. 55 → 60; × 27 → × 30; 1,800
9. \$84 → \$80; × 38 → × 40; \$3,200
10. 448 → 400; × 57 → × 60; 24,000
11. 637 → 600; × 42 → × 40; 24,000
12. \$275 → \$300; × 75 → × 80; \$24,000

Name ________________________

Modeling Multiplication

A model can be used to simplify multiplication.

This is a model for 12×26.

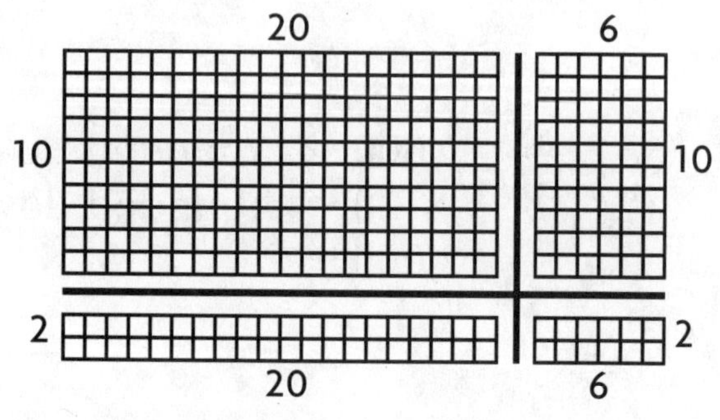

Multiply to find the number of squares in each rectangle. Then add the partial products.

$2 \times 6 =$ 12

$2 \times 20 =$ 40

$10 \times 6 =$ 60

$10 \times 20 =$ + 200

312

When you add the value of each rectangle in the grid, you find that the product of 12 and 26 is 312.

Multiply and solve by using the model and writing partial products.

1. $\begin{array}{r} 26 \\ \times 14 \\ \hline ? \end{array}$

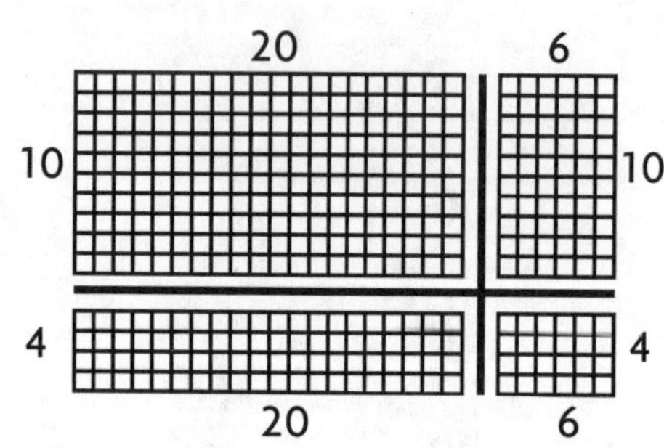

$4 \times 6 =$ **24**

$4 \times 20 =$ **80**

$10 \times 6 =$ **60**

$10 \times 20 =$ + **200**

364

2. $\begin{array}{r} 23 \\ \times 17 \\ \hline ? \end{array}$

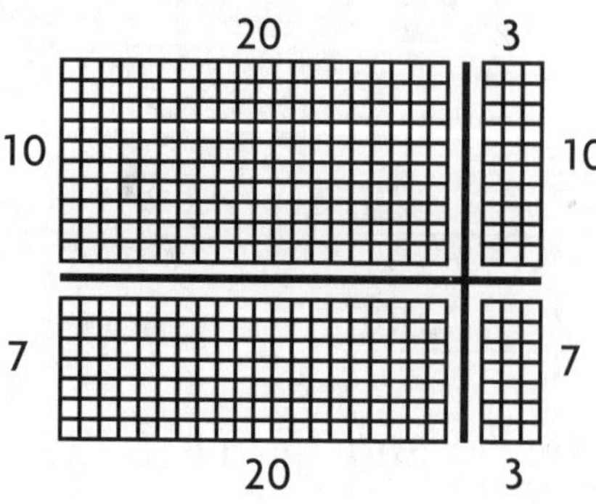

$7 \times 3 =$ **21**

$7 \times 20 =$ **140**

$10 \times 3 =$ **30**

$10 \times 20 =$ + **200**

391

Name ______________________

LESSON 17.4

Recording Multiplication

Solve 12 × 15 by using the partial-products method.

Step 1

```
  1(5)
× 1(2)
```

Step 2

```
  (1 5)
× 1 2)
```

Step 3

```
  1/5)
×(1/2
```

Step 4

```
 (1)5
×(1)2
```

Partial Products

		1	5
	×	1	2
2 × 5 =		1	0
2 × 10 =		2	0
10 × 5 =		5	0
10 × 10 =	1	0	0
product →	1	8	0

Step 5

Add partial products together.

10 + 20 + 50 + 100 = 180

The product of 12 and 15 is 180.

Multiply by using the partial-products method.

1.

		3	1
	×	1	7
7 × 1 =			**7**
7 × 30 =	**2**	**1**	**0**
10 × 1 =		**1**	**0**
10 × 30 =	**3**	**0**	**0**
product →	**5**	**2**	**7**

2.

			4	6
		×	2	8
8 × 6 =			**4**	**8**
8 × 40 =		**3**	**2**	**0**
20 × 6 =		**1**	**2**	**0**
20 × 40 =		**8**	**0**	**0**
product →	**1,**	**2**	**8**	**8**

3.

			7	9
		×	5	6
6 × 9 =			**5**	**4**
6 × 70 =		**4**	**2**	**0**
50 × 9 =		**4**	**5**	**0**
50 × 70 =	**3,**	**5**	**0**	**0**
product →	**4,**	**4**	**2**	**4**

4.

```
    82
   ×25
    10
   400
    40
 1,600
 2,050
```

5.

```
    63
   ×47
    21
   420
   120
 2,400
 2,961
```

6.

```
    92
   ×34
     8
   360
    60
 2,700
 3,128
```

Name ______________________________

LESSON 17.5

Practicing Multiplication

Find the product 28 × 136.

- Estimate by rounding. Since 30 × 100 is close to 3,000, your answer should be close to 3,000.
- Multiply.

```
   136
 × 28
 1,088 ← 136 × 8
 2,720 ← 136 × 20
 3,808
```

Remember:

136 × 8 is the same as

```
  8 × 6   =    48
  8 × 30  =   240
  8 × 100 =  +800
            1,088
```

- Compare the product with your estimate. Since 3,808 is close to your estimate of 3,000, it is a reasonable product.

Estimate and multiply.

1. 126 → 100
× 45 → × 50
5,000

		1	2	6
	×		4	5
		6	3	0
+	5,	0	4	0
	5,	6	7	0

2. 276 → 300
× 38 → × 40
12,000

		2	7	6
	×		3	8
	2,	2	0	8
+	8,	2	8	0
1	0,	4	8	8

3. 737 → 700
× 47 → × 50
35,000

		7	3	7
	×		4	7
	5,	1	5	9
+ 2	9,	4	8	0
3	4,	6	3	9

4. 545 → 500
× 24 → × 20
10,000

		5	4	5
	×		2	4
	2,	1	8	0
+ 1	0,	9	0	0
1	3,	0	8	0

Name ______________________________

LESSON 18.1

Modeling Division

Divide 92 into 4 equal groups.

Write. 92 ÷ 4

Step 1

Show 92 as 9 tens and 2 ones.

Draw 4 circles to show 4 groups.

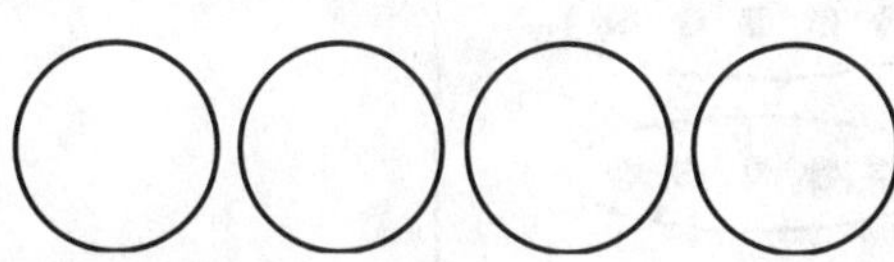

Step 2

Begin by dividing the 9 tens. Place an equal number of tens into each circle.

$$\begin{array}{rl} 2 & \text{— 2 tens in each group} \\ 4\overline{)\mathbf{92}} & \\ -8 & \text{— 8 tens used} \\ \hline 1 & \text{— 1 ten left} \end{array}$$

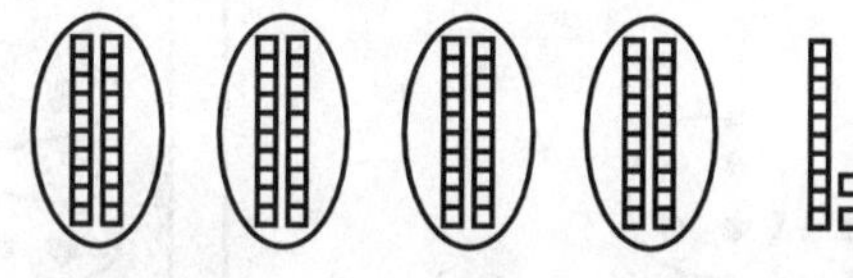

Step 3

Regroup the 1 ten left over into ones. Now there are 12 ones.

$$\begin{array}{rl} 2 & \\ 4\overline{)92} & \text{bring} \\ -8\downarrow & \text{ones} \\ \hline 12 & \text{—down} \end{array}$$

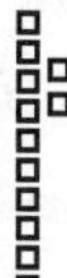    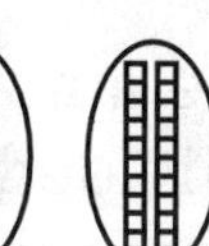

Step 4

Divide the 12 ones. Place an equal number of ones into each circle.

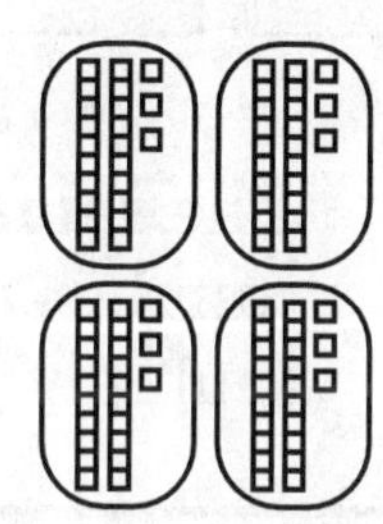

$$\begin{array}{rl} 23 & \text{—3 ones in each group} \\ 4\overline{)92} & \\ -8 & \\ \hline \mathbf{12} & \\ -12 & \text{—12 ones used} \\ \hline 0 & \text{—0 ones left} \end{array}$$

92 ÷ 4 = 23

Use base-ten blocks to model each problem. Record the numbers as you complete each step.

1.

	2	**4**
2)	4	8
−	**4**	
	0	**8**
−		**8**
		0

2.

	3	**2**
3)	9	6
−	**9**	
	0	**6**
−		**6**
		0

3.

	1	**3**
4)	5	2
−	**4**	
	1	**2**
−	**1**	**2**
		0

4.

	1	**7**
2)	3	4
−	**2**	
	1	**4**
−	**1**	**4**
		0

5.

	1	**3**
5)	6	5
−	**5**	
	1	**5**
−	**1**	**5**
		0

Name ______________________________

LESSON 18.2

Dividing with Remainders

What is 23 ÷ 5?

The drawings in the boxes below show two different ways to think about this division problem.

Example A

Think: If you divide 23 into 5 equal groups, how many are in each group?

5 × [?] = 23

There are 5 groups of 4, with 3 left over.

Example B

Think: How many groups of 5 can you make from 23?

[?] × 5 = 23

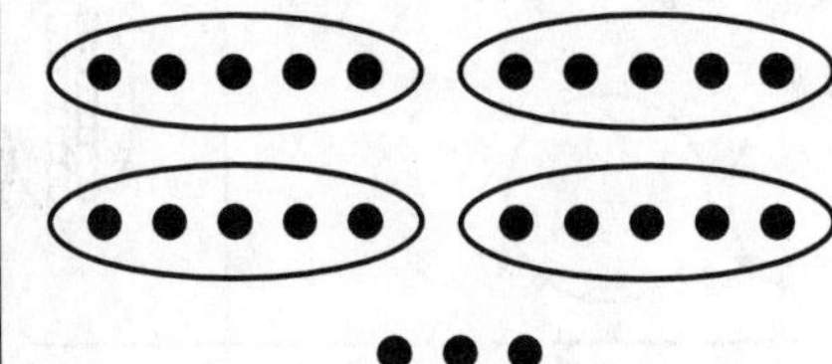

There are 4 groups of 5, with 3 left over.

23 ÷ 5 = 4 r3

$$\begin{array}{r} 4\text{ r}3 \\ 5\overline{)23} \\ -20 \\ \hline 3 \end{array}$$

The **remainder** is the amount left over when a number cannot be divided evenly. You can multiply and subtract to find the remainder.

Complete Exercises 1–5 to find the remainder.

1. $\begin{array}{r} 6\text{ r}\boxed{2} \\ 5\overline{)32} \\ -30 \\ \hline \boxed{2} \end{array}$

2. $\begin{array}{r} 4\text{ r}\boxed{3} \\ 7\overline{)31} \\ -28 \\ \hline \boxed{3} \end{array}$

3. $\begin{array}{r} 9\text{ r}\boxed{3} \\ 4\overline{)39} \\ -\boxed{36} \\ \hline \boxed{3} \end{array}$

4. $\begin{array}{r} 6\text{ r}\boxed{1} \\ 3\overline{)19} \\ -\boxed{18} \\ \hline \boxed{1} \end{array}$

5. $\begin{array}{r} 6\text{ r}\boxed{8} \\ 9\overline{)62} \\ -\boxed{54} \\ \hline \boxed{8} \end{array}$

Find the quotient.

6. $\begin{array}{r} 6\text{ r}2 \\ 5\overline{)32} \\ -30 \\ \hline 2 \end{array}$

7. $\begin{array}{r} 3\text{ r}4 \\ 5\overline{)19} \\ -15 \\ \hline 4 \end{array}$

8. $\begin{array}{r} 7\text{ r}1 \\ 3\overline{)22} \\ -21 \\ \hline 1 \end{array}$

9. $\begin{array}{r} 8\text{ r}1 \\ 2\overline{)17} \\ -16 \\ \hline 1 \end{array}$

Name ______________________________

LESSON 18.3

Division Procedures

What is 47 ÷ 3? Remember these steps:

D ivide **M** ultiply **S** ubtract **C** ompare and **B** ring down

1 3)47 −3 1	Divide the tens. Multiply. Subtract. Compare.	1 3)4 $3 \times 1 = 3$ $4 - 3 = 1$ $1 < 3$
1 3)47 −3↓ 17	Bring down?	Yes. Bring the 7 down to make 17.
15 3)47 −3 17 −15 2	Divide the ones. Multiply. Subtract. Compare.	5 3)17 $3 \times 5 = 15$ $17 - 15 = 2$ $2 < 3$
15 r2 3)47 −3 17 −15 2	Bring down?	Nothing is left to bring down. If the number left over is smaller than the divisor, record it as the remainder.

Solve. Use the D, M, S, C, B steps shown above.

1. 13
5)65
−5
15
−15
0

2. 17 r1
3)52
−3
22
−21
1

3. 25 r1
2)51
−4
11
−10
1

4. 22 r3
4)91
−8
11
−8
3

Name ____________________

LESSON 18.4

Placing the First Digit in the Quotient

When you begin a division problem, you must first decide where to place the first digit of the quotient.

Example A

tens ones	
3)71: 2 over tens; − 6 ↓; 11	**Divide the tens.** **Multiply.** $3 \times 2 = 6$ **Subtract.** $7 - 6 = 1$ **Compare.** $1 < 3$ **Bring down?** Yes. Bring down 1 to make 11.
3)71: 23 r2; − 6; 11; − 9; 2	**Divide the ones.** **Multiply.** $3 \times 3 = 9$ **Subtract.** $11 - 9 = 2$ **Compare.** $2 < 3$ **Bring down?** Nothing is left. If there is a remainder, record it.

Example B

tens ones	
4)39	$3 < 4$, so ask yourself how many 4's are in 39.
4)39: 9 r3; − 36; 3	**Divide the ones.** **Multiply.** **Subtract.** **Compare.** **Bring down?** Nothing is left. If there is a remainder, record it.

Write an **x** where the first digit in the quotient should be placed.

1. 5)34: x over ones; − 30; 4
2. 3)74: x over tens; − 6 ↓; 14
3. 2)19: x over ones; − 18; 1
4. 4)53: x over tens; − 4 ↓; 13
5. 5)83: x over tens; − 5 ↓; 33

Find the quotient.

6. 3)29: 9 r2; − 27; 2
7. 2)47: 23 r1; − 4; 07; − 6; 1
8. 7)19: 2 r5; − 14; 5
9. 5)75: 15; − 5; 25; − 25; 0

Name ______________________________

Problem-Solving Strategy

Guess and Check

There are 87 marbles. After the marbles are divided evenly into bags, there are 2 marbles left over. There are fewer than 10 bags. How many bags are there? How many marbles are in each bag?

Guess:	3 Bags?	4 Bags?	5 Bags?
Check: Remember, there must be 2 marbles remaining.	29 3)87 −6 27 −27 0 No remainder	21 r3 4)87 −8 07 − 4 3 Remainder of 3	17 r2 5)87 −5 37 −35 2 Remainder of 2! You found the answer.

There are 5 bags with 17 marbles in each bag, and 2 marbles are left over.

There are 82 cookies. When the cookies are divided evenly on plates, there are 4 cookies left over. There are fewer than 10 plates. How many plates are there? How many cookies are there on each plate?

Hint: Guess numbers between 0 and 9. You may need to guess more than three times.

Guess:	____ Plates?	____ Plates?	____ Plates?
Check: Remember, there must be 4 cookies remaining.	**6 plates with 13 cookies on each**		

Name ______________________________

Dividing Three-Digit Numbers

These examples show how to divide a three-digit number.

Example A

1 3)529 −3↓ 2	**D**ivide the hundreds. **M**ultiply. **S**ubtract. **C**ompare. **B**ring down?
17 3)529 −3 **22** −21↓ 19	**D**ivide the tens. **M**ultiply. **S**ubtract. **C**ompare. **B**ring down?
176 r1 3)529 −3 22 −21 **19** −18 1	**D**ivide the ones. **M**ultiply. **S**ubtract. **C**ompare. **B**ring down? If there is a remainder, record it.

Example B

6)**1**59	There are not enough hundreds to divide.
2 6)**15**9 − 12↓ 39	**D**ivide the tens. **M**ultiply. **S**ubtract. **C**ompare. **B**ring down?
26 r3 6)159 −12 **39** −36 3	**D**ivide the ones. **M**ultiply. **S**ubtract. **C**ompare. **B**ring down? If there is a remainder, record it.

Write an **x** where the first digit in the quotient should be placed.

1. 5)349 — x over 4; −30↓; 49
2. 3)745 — x over 7; −6↓; 14
3. 2)198 — x over 9; −18↓; 11
4. 4)532 — x over 5; −4↓; 13

Find the quotient.

5. 5)349 = 69 r4; −30; 49; −45; 4
6. 3)745 = 248 r1; −6; 14; −12; 25; −24; 1
7. 2)198 = 99; −18; 18; −18; 0
8. 4)532 = 133; −4; 13; −12; 12; −12; 0

Name ______________________________

Practicing Division

In some division problems, you must be careful to place a zero in the tens place or the ones place.

Example A

1 4)429 −4↓ 02	Divide the hundreds. Multiply. Subtract. Compare. Bring down?
10 4)429 −4↓ **02** − 0↓ 29	There are not enough tens to divide. 2 < 4 Write 0. Multiply. Subtract. Compare. Bring down?
107 r1 4)429 −4↓ 02 − 0↓ **29** −28 1	Divide the ones. Multiply. Subtract. Compare. Bring down? If there is a remainder, record it.

Example B

2 3)721 −6↓ 12	Divide the hundreds. Multiply. Subtract. Compare. Bring down?
24 3)721 −6↓ **12** −12↓ 01	Divide the tens. Multiply. Subtract. Compare. Bring down?
240 r1 3)721 −6↓ 12 −12↓ **01** − 0 1	There are not enough ones to divide. 1 < 3 Write 0. Multiply. Subtract. Compare. Bring down? If there is a remainder, record it.

Find the quotient.

1.

	1	0	9 r2
3)	3	2	9
−	3		
	0	2	
−		0	
		2	9
	−	2	7
			2

2.

	2	3	0 r3
4)	9	2	3
−	8		
	1	2	
−	1	2	
		0	3
		−	0
			3

3.

	1	0	8 r1
5)	5	4	1
−	5		
	0	4	
−		0	
		4	1
	−	4	0
			1

4.

	1	2	0 r4
6)	7	2	4
−	6		
	1	2	
−	1	2	
		0	4
		−	0
			4

Name ____________________

LESSON 19.1

Division Patterns to Estimate

Division Patterns

You can find how many fives are in 2,000 by following these steps.

- Rewrite each problem as a multiplication fact.
- Circle all the numbers in the basic fact to see how many zeros to write in the quotient.
- Write the quotient. If there is a zero in the basic fact, the quotient has one less zero than the dividend. Otherwise the quotient has the same number of zeros as the dividend. You can estimate a quotient by using numbers that are easy to divide.

$20 \div 5 = 4$ (5)×(4)=(20)

$200 \div 5 = 40$ (5)×(4)**0** =(20)**0**

$2{,}000 \div 5 = ?$ (5)×(4)**00** =(20)**00**

So, $2{,}000 \div 5 = 400$.

Find the quotient. Use a basic fact and a pattern of zeros to help you.

1. $45 \div 9 =$ **5**
   $450 \div 9 =$ **50**
   $4{,}500 \div 9 =$ **500**
2. $21 \div 3 =$ **7**
   $210 \div 3 =$ **70**
   $2{,}100 \div 3 =$ **700**
3. $81 \div 9 =$ **9**
   $810 \div 9 =$ **90**
   $8{,}100 \div 9 =$ **900**
4. $36 \div 6 =$ **6**
   $360 \div 6 =$ **60**
   $3{,}600 \div 6 =$ **600**
5. $72 \div 8 =$ **9**
   $720 \div 8 =$ **90**
   $7{,}200 \div 8 =$ **900**
6. $12 \div 4 =$ **3**
   $120 \div 4 =$ **30**
   $1{,}200 \div 4 =$ **300**

Estimate the quotient.

7. $278 \div 4 = n$
   278 is close to **280**.
   280 $\div 4 =$ **70**
8. $142 \div 7 = n$
   142 is close to **140**.
   140 $\div 7 =$ **20**
9. $357 \div 9 = n$
   357 is close to **360**.
   360 $\div 9 =$ **40**

Name ______________________________

Zeros in Division

In some division problems, you need to place a zero in the tens place or the ones place of the quotient.

Zero in tens place

	H	T	O
	2	0	8
4)	8	3	2
−	8		
	0	3	
−		0	
		3	2
	−	3	2
			0

$4\overline{)8}$	$4 \times 2 = 8$ Write 2 in quotient.
$4\overline{)3}$	$4 \times 0 = 0$ **Write 0 in quotient.**
$4\overline{)32}$	$4 \times 8 = 32$ Write 8 in quotient.

Zero in ones place

	H	T	O	
	2	5	0	r2
3)	7	5	2	
−	6			
	1	5		
−	1	5		
		0	2	
		−	0	
			2	

$3\overline{)7}$	$3 \times 2 = 6$ Write 2 in quotient.
$3\overline{)15}$	$3 \times 5 = 15$ Write 5 in quotient.
$3\overline{)2}$	$3 \times 0 = 0$ **Write 0 in quotient.**

Find the quotient.

1.

	2	0	4	r2
3)	6	1	4	
−	6			
	0	1		
−		0		
		1	4	
	−	1	2	
			2	

2.

	1	0	7
4)	4	2	8
−	4		
	0	2	
−		0	
		2	8
	−	2	8
			0

3.

	4	0	9	r1
2)	8	1	9	
−	8			
	0	1		
−		0		
		1	9	
	−	1	8	
			1	

4.

	1	3	0	r3
7)	9	1	3	
−	7			
	2	1		
−	2	1		
		0	3	
	−		0	
			3	

5.

	2	8	0
3)	8	4	0
−	6		
	2	4	
−	2	4	
		0	0
	−		0
			0

6.

	1	3	0	r2
5)	6	5	2	
−	5			
	1	5		
−	1	5		
		0	2	
	−		0	
			2	

Name ___________________________

LESSON 19.3

Practicing Division

You can use play money to practice division with money amounts.

Divide. $5.26 ÷ 2

Step 1	Step 2
Divide the $1 bills into 2 equal groups.   $\begin{array}{r} \$2. \\ 2\overline{)\$5.26} \\ -4 \\ \hline 1 \end{array}$	Regroup the one $1 bill left over into 10 dimes. 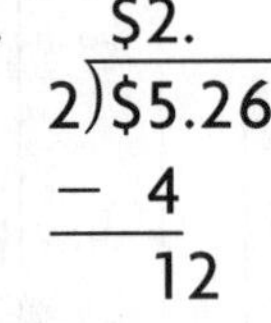   $\begin{array}{r} \$2. \\ 2\overline{)\$5.26} \\ -4 \\ \hline 12 \end{array}$
Step 3	**Step 4**
Divide the dimes into 2 equal groups.  $\begin{array}{r} \$2.6 \\ 2\overline{)\$5.26} \\ -4 \\ \hline 12 \\ -12 \\ \hline \end{array}$	Divide the pennies into 2 equal groups.   $\begin{array}{r} \$2.63 \\ 2\overline{)\$5.26} \\ -4 \\ \hline 12 \\ -12 \\ \hline 06 \\ -6 \\ \hline 0 \end{array}$

Find the quotient. Use play money to help you. Line up the dollar sign and the decimal point.

1. $\begin{array}{r} \$1.81 \\ 4\overline{)\$7.24} \\ -4 \\ \hline 32 \\ -32 \\ \hline 04 \\ -4 \\ \hline 0 \end{array}$

2. $\begin{array}{r} \$2.06 \\ 3\overline{)\$6.18} \\ -6 \\ \hline 01 \\ -0 \\ \hline 18 \\ -18 \\ \hline 0 \end{array}$

3. $\begin{array}{r} \$0.52 \\ 5\overline{)\$2.60} \\ -0 \\ \hline -25 \\ \hline 10 \\ -10 \\ \hline 0 \end{array}$

4. $\begin{array}{r} \$2.64 \\ 2\overline{)\$5.28} \\ -4 \\ \hline 12 \\ -12 \\ \hline 08 \\ -8 \\ \hline 0 \end{array}$

Name ______________________________

LESSON 19.4

Meaning of the Remainder

The way you interpret the remainder in a division problem depends on the situation.

Jeff took 23 photographs during his trip. He can fit 4 photographs on each page of his album. How many pages will it take to fit all 23 photographs?

Since 5 pages are not enough to fit all the photographs, Jeff needs to use 6 pages.

Julia is making bouquets of flowers. She has 20 flowers in all. If she puts 6 flowers in each bouquet, how many bouquets can she make?

Julia can make 3 bouquets. The remainder is not enough to make another bouquet.

Solve. Tell how you interpret the meaning of the remainder. **Explanations will vary.**

1. Sam is packaging cookies for a bake sale. He has 26 cookies. He puts 3 cookies in each bag. How many bags of cookies does he have?

   8 bags;
   drop the remainder.

2. A group of 42 students are going on a field trip in vans that hold 9 students each. How many vans are needed to transport all the students?

   5 vans; round to the next
   greater whole number.

3. Becky made 32 clay beads. How many necklaces can she make if she uses 5 beads for each necklace?

   6 necklaces; drop the
   remainder.

4. Chris wants to buy cupcakes for 14 people. The cupcakes come 6 to a package. How many packages of cupcakes does Chris need to buy so that each person can have 2 cupcakes?

   buy 5 packages; round to the
   next greater whole number.

Name ______________________________

LESSON 19.4

Problem Solving

Account for All Possibilities

In some situations, you may want to divide a number into groups so that there are exactly the same number in each group.

Mary wants to give the same number of marbles to each of her 3 friends. She has a bag of 23 marbles. How many marbles can Mary give to each friend?

She can give each friend 7 marbles. There are 2 marbles left over.

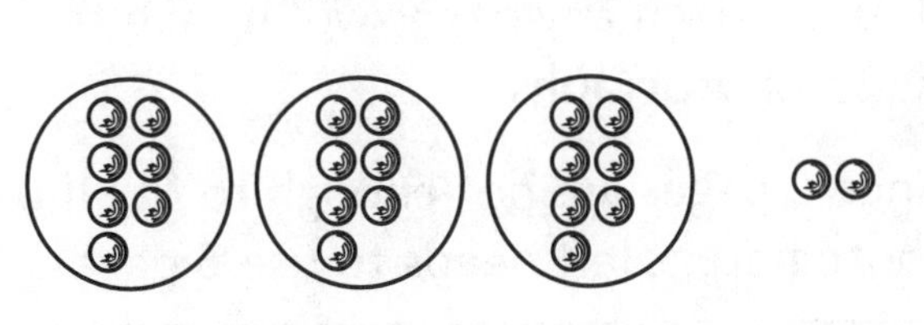

In other situations, you may use the remainder to make one or more of the groups larger.

Megan is making 26 cards for her classmates. She has 4 days to make the cards. How many cards should Megan make each day?

She can make 6 cards on 2 of the days, and 7 cards on 2 of the days.

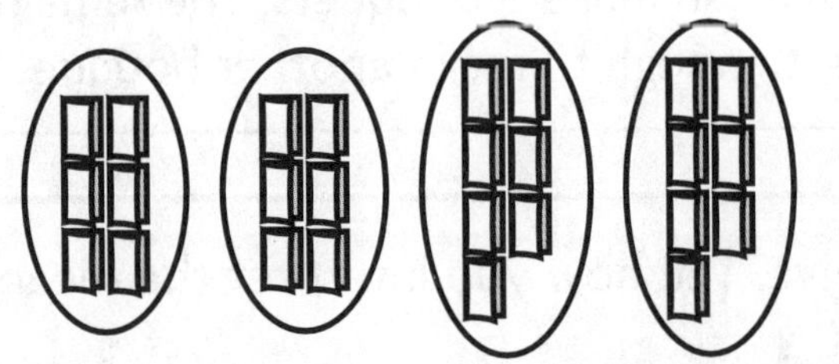

Solve. Account for all the possibilities.

1. Chen wants to divide his 136 shells as evenly as possible into 3 boxes. How many shells can he put in each box?

   45 shells in 2 boxes; 46 shells in 1 box

2. Joe, Jack, and Henry want to divide $4.35 so that they each get the same amount of money. How much money will each boy get? Will there be any money left over?

   $1.45; no

3. Nan has borrowed a 120-page book that is due at the library in 7 days. If she wants to finish the book in 7 days, how many pages should she read each day?

   17 pages on 6 days; 18 pages on 1 day

4. There are 4 students who are working together to write 50 math problems for their class to solve. How many problems should each student write?

   12 problems for 2 students; 13 problems for 2 students

Name ______________________________

Finding the Average

An **average** is one number that best represents all the numbers in a set.

To find the average:

Step 1 Add the numbers.

Step 2 Count the number of addends.

Step 3 Divide the sum by the number of addends.

Examples:

Find the average of 3, 4, 6, and 7.

Step 1 $3 + 4 + 6 + 7 = 20$

Step 2 There are 4 addends.

Step 3 $20 \div 4 = 5$

The average is 5.

Find the average of 5, 7, 8, 9, and 11.

Step 1 $5 + 7 + 8 + 9 + 11 = 40$

Step 2 There are 5 addends.

Step 3 $40 \div 5 = 8$

The average is 8.

Find the average of each set of numbers.

1. 5, 6, and 10

Step 1 $5 + 6 + 10 =$ __21__

Step 2 There are __3__ addends.

Step 3 __21__ $\div 3 =$ __7__

2. 2, 4, 5, 6, and 8

Step 1 $2 + 4 + 5 + 6 + 8 =$ __25__

Step 2 There are __5__ addends.

Step 3 __25__ $\div 5 =$ __5__

3. 4, 6, 8, and 10

Step 1 __4__ + __6__ + __8__ + __10__ = __28__

Step 2 There are __4__ addends.

Step 3 __28__ ÷ __4__ = __7__

4. 6, 8, and 10

Step 1 __6__ + __8__ + __10__ = __24__

Step 2 There are __3__ addends.

Step 3 __24__ ÷ __3__ = __8__

Find the average.

5. 4, 5, 5, 8, 8

__6__

6. 1, 2, 4, 5

__3__

7. 4, 7, 10

__7__

Name ____________________

Choosing the Operation

Add	Joining groups of different sizes	There are 24 students in Mr. Clark's class and 26 students in Ms. Young's class. How many students are there in both classes? $24 + 26 = 50$ 50 students
Subtract	Taking away or comparing groups	How many more students are in Ms. Young's class than in Mr. Clark's class? $26 - 24 = 2$ 2 more students
Multiply	Joining equal-sized groups	Mr. Clark gave 3 pencils to each of his 24 students. How many pencils in all did Mr. Clark give his students? $24 \times 3 = 72$ 72 pencils
Divide	Finding how many in each group or how many groups	Mr. Clark divided his class of 24 students into 4 equal groups to work on a science project. How many students were in each group? $24 \div 4 = 6$ 6 students

Name the operation you use. Write add, subtract, multiply, or divide. Then solve.

1. Maria used 3 rolls of film on her vacation. Each roll had 36 exposures. How many photographs did she take?

   multiply; 108 photographs

2. Maria bought a photograph album for $12.95. What was her change from a $20 bill?

   subtract; $7.05

3. Maria put 6 photographs on each page of her album. How many pages did she use for her 84 photographs of the Rocky Mountains?

   divide; 14 pages

4. During a 4-day hiking trip, Maria hiked 5 km, 7 km, 9 km, and 10 km. What was the total length of her hike?

   add; 31 km

Name ______________________________

Fractions: Part of a Whole

A **fraction** is a number that names a part of a whole.

	$\frac{\text{shaded parts}}{\text{total parts}} = \frac{2}{3}$	**Read:** two thirds two out of three two divided by three
	$\frac{\text{shaded parts}}{\text{total parts}} = \frac{4}{6}$	**Read:** four sixths four out of six four divided by six

Write a fraction for each model.

1. 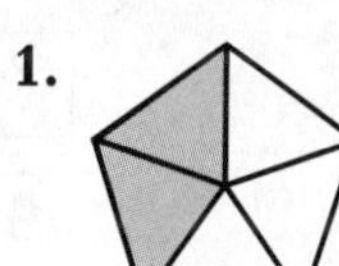 $\frac{\text{shaded parts}}{\text{total parts}} = \frac{2}{5}$

2. 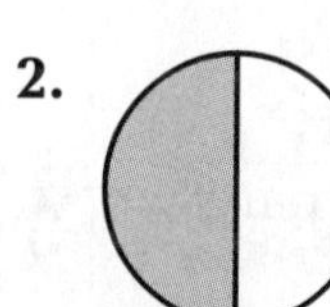 $\frac{\text{shaded parts}}{\text{total parts}} = \frac{1}{2}$

3. 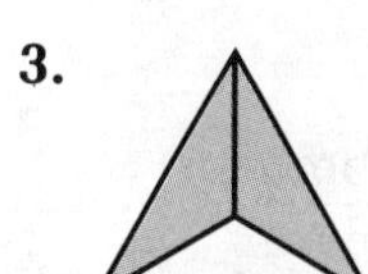 $\frac{\text{shaded parts}}{\text{total parts}} = \frac{2}{3}$

4. 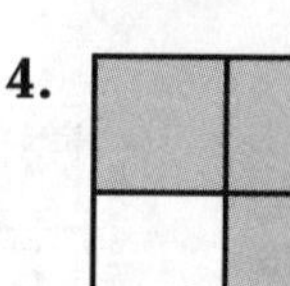 $\frac{\text{shaded parts}}{\text{total parts}} = \frac{3}{4}$

5. 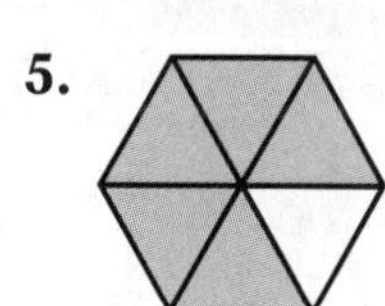 $\frac{\text{shaded parts}}{\text{total parts}} = \frac{5}{6}$

6. 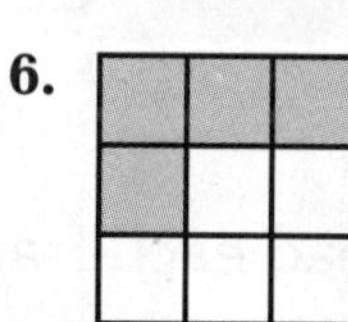 $\frac{\text{shaded parts}}{\text{total parts}} = \frac{4}{9}$

Write three ways you could read each fraction.

7. $\frac{3}{4}$ three fourths, three out of four, three divided by four

8. $\frac{1}{2}$ one half, one out of two, one divided by two

9. $\frac{2}{3}$ two thirds, two out of three, two divided by three

10. $\frac{1}{5}$ one fifth, one out of five, one divided by five

11. $\frac{2}{8}$ two eighths, two out of eight, two divided by eight

12. $\frac{8}{9}$ eight ninths, eight out of nine, eight divided by nine

Name ______________________

LESSON 20.2

Fractions: Part of a Group

A **fraction** can also name a part of a group.

	$\frac{\text{parts shaded}}{\text{number of equal parts}} = \frac{1}{2}$	**Read:** one half one out of two one divided by two
	$\frac{\text{parts with filled glasses}}{\text{number of equal parts}} = \frac{4}{5}$	**Read:** four fifths four out of five four divided by five

Write a fraction for each model.

1. $\frac{\text{parts with cats}}{\text{number of equal parts}} = \frac{2}{3}$

2. $\frac{\text{parts with stars}}{\text{number of equal parts}} = \frac{7}{9}$

3. $\frac{\text{part with baseballs}}{\text{number of equal parts}} = \frac{1}{2}$

4. $\frac{\text{parts with open umbrellas}}{\text{number of equal parts}} = \frac{3}{4}$

Write three ways you could read each fraction.

5. $\frac{1}{10}$ one tenth, one out of ten, one divided by ten

6. $\frac{3}{5}$ three fifths, three out of five, three divided by five

7. $\frac{2}{3}$ two thirds, two out of three, two divided by three

8. $\frac{2}{9}$ two ninths, two out of nine, two divided by nine

9. $\frac{6}{7}$ six sevenths, six out of seven, six divided by seven

10. $\frac{5}{10}$ five tenths, five out of ten, five divided by ten

Name ______________________________

LESSON 20.3

Equivalent Fractions

1			
$\frac{1}{4}$			
$\frac{1}{8}$		$\frac{1}{8}$	
$\frac{1}{16}$	$\frac{1}{16}$	$\frac{1}{16}$	$\frac{1}{16}$

Fractions that name the same amount are called **equivalent fractions.** $\frac{1}{4}$, $\frac{2}{8}$, and $\frac{4}{16}$ are different names for the same number. So, $\frac{1}{4} = \frac{2}{8} = \frac{4}{16}$, which makes them equivalent fractions.

Complete to find the equivalent fraction.

1. $\frac{2}{3} = \frac{\boxed{4}}{6}$

2. $\frac{1}{2} = \frac{\boxed{4}}{8}$

3.

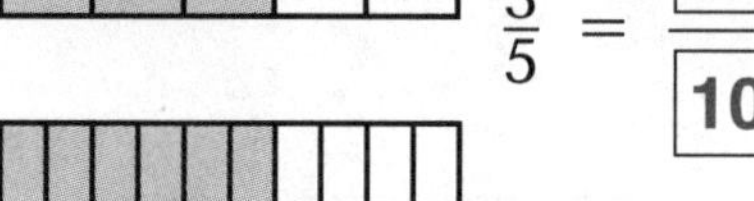

$\frac{3}{5} = \frac{\boxed{6}}{\boxed{10}}$

4.

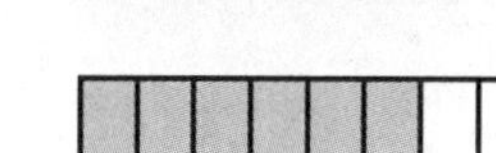

$\frac{3}{4} = \frac{\boxed{6}}{\boxed{8}}$

Color the correct number of parts to show the equivalent fraction. Then write the equivalent fraction.

5.

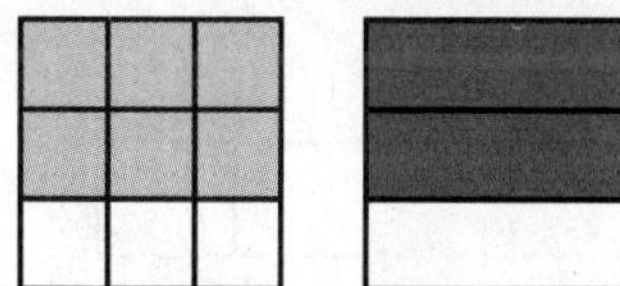

$\frac{6}{9} = \frac{\boxed{2}}{\boxed{3}}$

6.

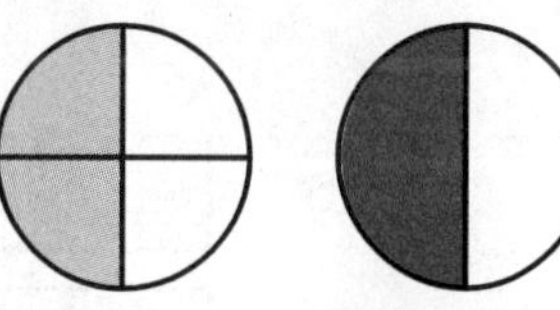

$\frac{2}{4} = \frac{\boxed{1}}{\boxed{2}}$

7.

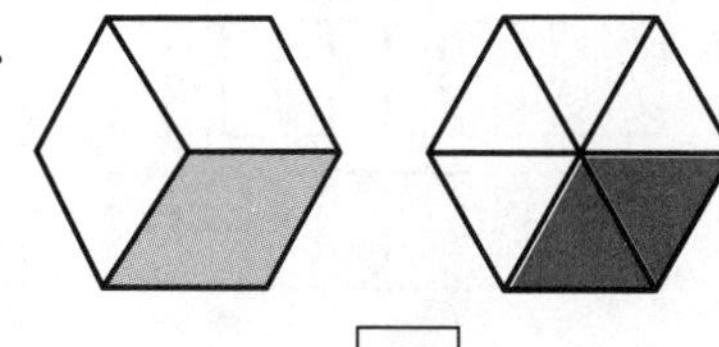

$\frac{1}{3} = \frac{\boxed{2}}{\boxed{6}}$

8.

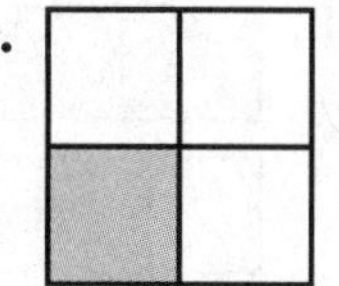

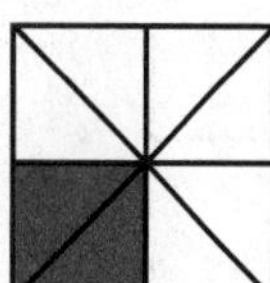

$\frac{1}{4} = \frac{\boxed{2}}{\boxed{8}}$

Name ______________________

Comparing and Ordering Fractions

You can compare fractions that have the same denominators.

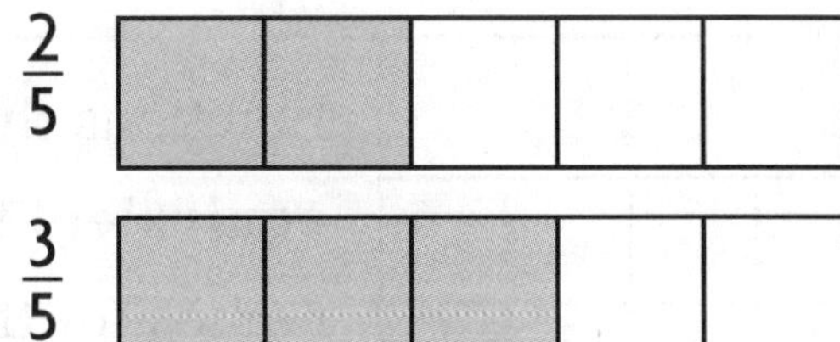

Compare $\frac{2}{5}$ and $\frac{3}{5}$.
Compare the shaded areas in the fraction models.
$2 < 3$, so $\frac{2}{5} < \frac{3}{5}$.

You can also compare fractions that have different denominators.

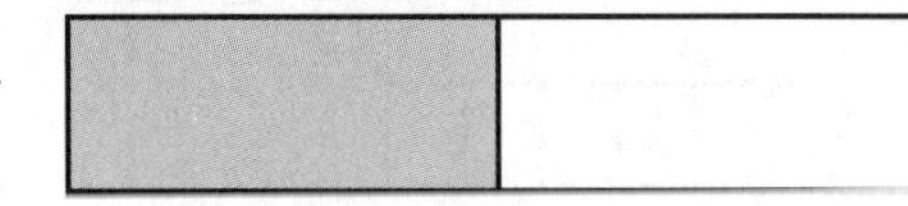

Compare $\frac{2}{3}$ and $\frac{1}{2}$.
Compare the shaded areas in the fraction models.
Since $\frac{2}{3}$ has a larger shaded area, $\frac{2}{3} > \frac{1}{2}$.

Compare the fraction models. Write <, >, or =.

1. 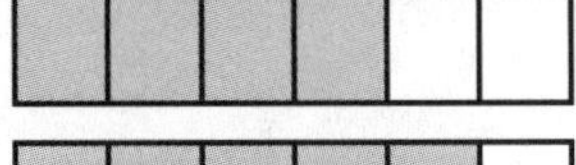

$\frac{4}{6}$ (<) $\frac{5}{6}$

2. 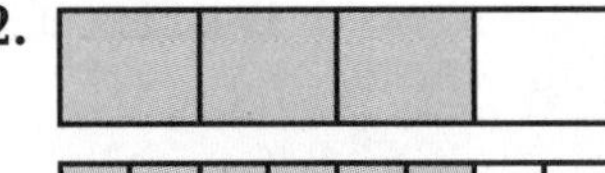

$\frac{3}{4}$ (=) $\frac{6}{8}$

3. 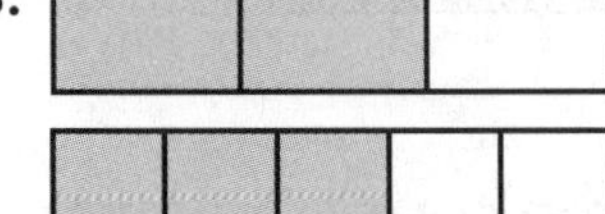

$\frac{2}{3}$ (>) $\frac{3}{5}$

4. 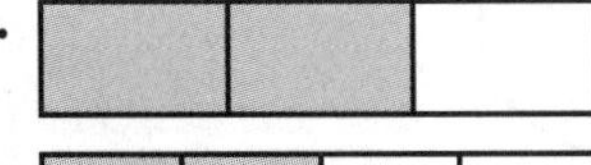

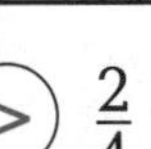

$\frac{2}{3}$ (>) $\frac{2}{4}$

5. 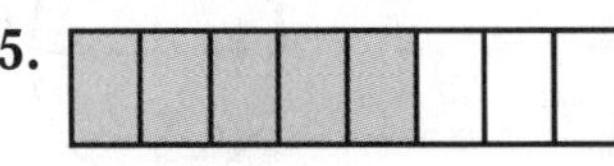

$\frac{5}{8}$ (<) $\frac{6}{8}$

6. 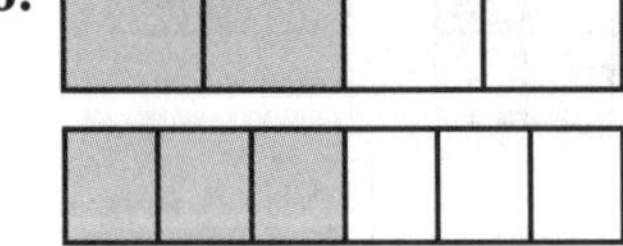

$\frac{2}{4}$ (=) $\frac{3}{6}$

7. 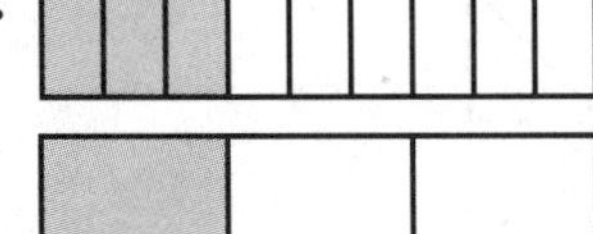

$\frac{3}{9}$ (=) $\frac{1}{3}$

8. 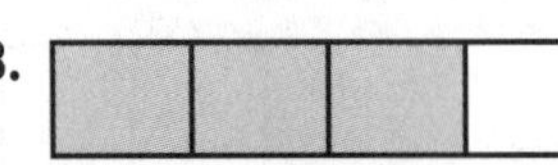

$\frac{3}{4}$ (>) $\frac{3}{5}$

9. 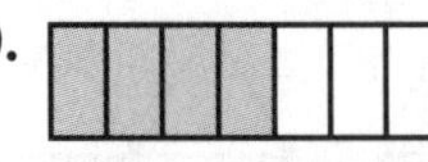

$\frac{4}{10}$ (<) $\frac{5}{8}$

Name ______________________

Problem-Solving Strategy

Make a Model

Beth made a snack mix for the class party. She used $\frac{1}{2}$ lb of peanuts, $\frac{2}{3}$ lb of pretzels, and $\frac{1}{4}$ lb of dried fruit. List the ingredients in order from greatest to least amount.

You can make a model to help solve the problem. Use fraction bars to model the fractions. Line up the fraction bars so you can tell which bar is the largest and which bar is the smallest.

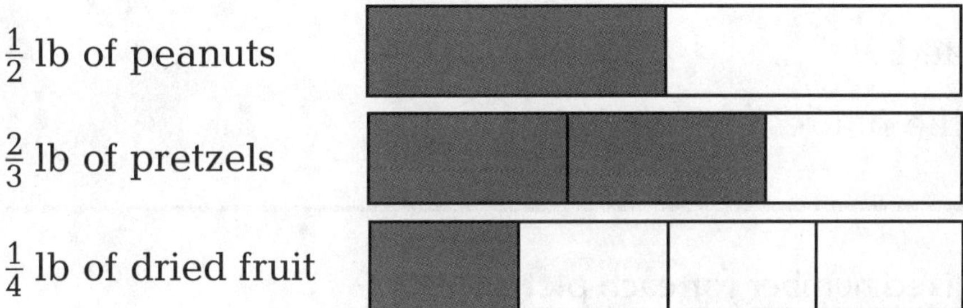

The bar for pretzels is the largest and the bar for dried fruit is the smallest. So, the ingredients in order from greatest to least is pretzels, peanuts, and dried fruit.

Make a model to solve. **Check students' models.**

1. The amusement park has a colorful sidewalk. Starting with the first square, every third square is orange, and every fifth square is green. Draw a model of the first 12 squares of the sidewalk in the space below. What fraction of the first 12 squares are either orange or green?

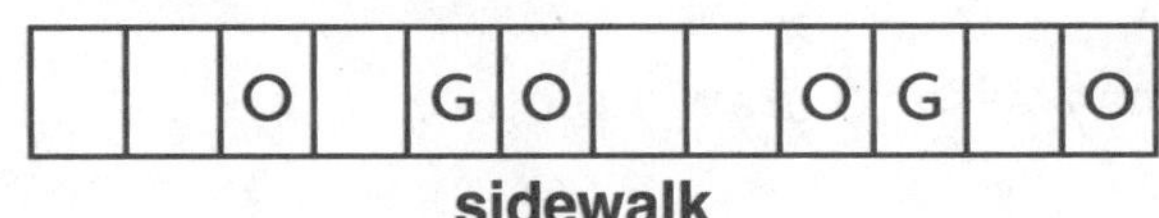

sidewalk

$\frac{6}{12}$ of the first 12 squares are either orange or green.

2. Patti made punch from $\frac{1}{2}$ gallon of grape juice, $\frac{2}{3}$ gallon of kiwi juice, and $\frac{5}{8}$ gallon of papaya juice. Draw a model for each in the space below. Then list the ingredients in order from greatest to least.

grape juice

kiwi juice

papaya juice

kiwi juice, papaya juice, grape juice

Name ______________________________

LESSON 20.5

Mixed Numbers

A **mixed number** is made up of a whole number and a fraction.

These fraction bars represent a mixed number.

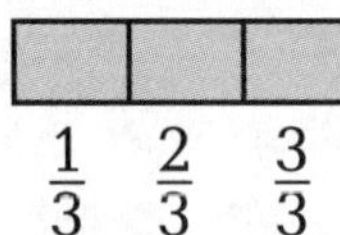
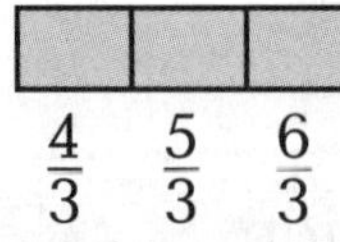
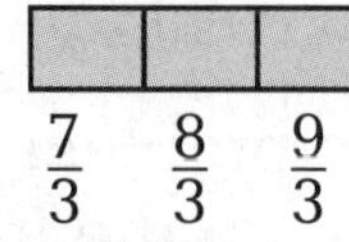

$\frac{1}{3}$ $\frac{2}{3}$ $\frac{3}{3}$ $\frac{4}{3}$ $\frac{5}{3}$ $\frac{6}{3}$ $\frac{7}{3}$ $\frac{8}{3}$ $\frac{9}{3}$ $\frac{10}{3}$ $\frac{11}{3}$

There are 3 whole figures shaded.

The last figure is $\frac{2}{3}$ shaded.

So, $3\frac{2}{3}$ figures are shaded.

Or, you can say $\frac{11}{3}$ of the figures are shaded.

Write a fraction and a mixed number for each picture.

1.

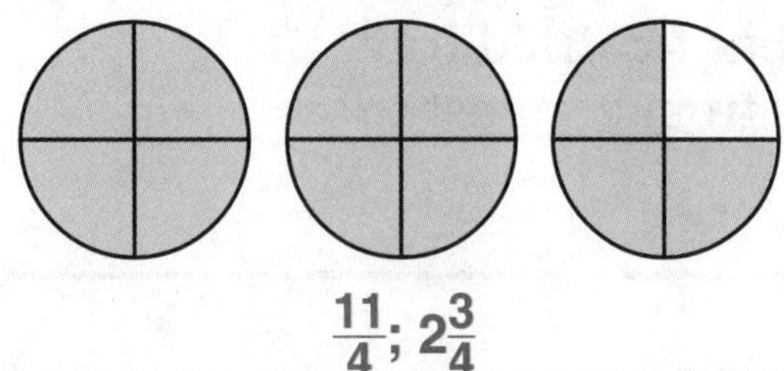

$\frac{11}{4}$; $2\frac{3}{4}$

2.

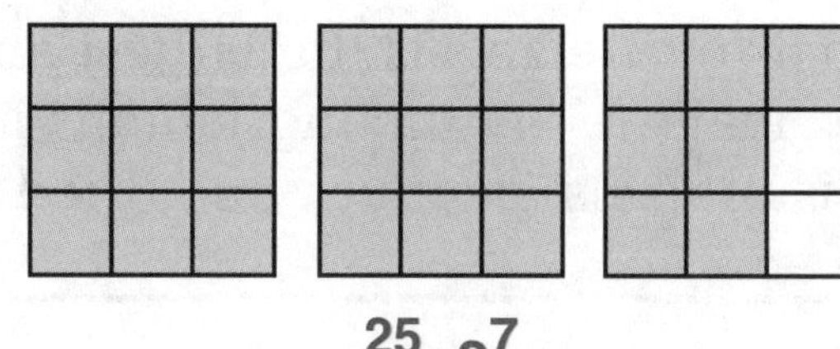

$\frac{25}{9}$; $2\frac{7}{9}$

3.

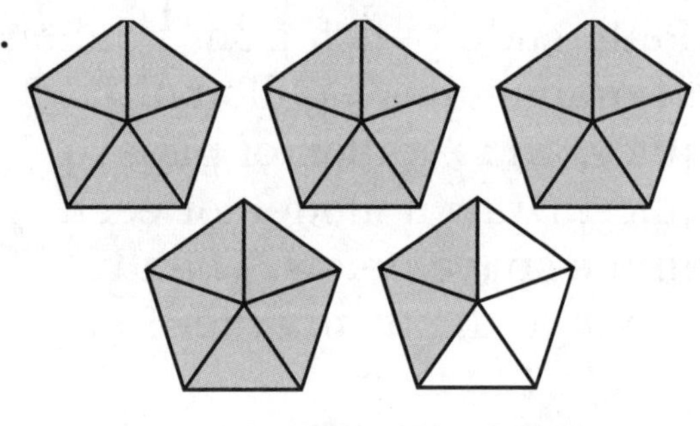
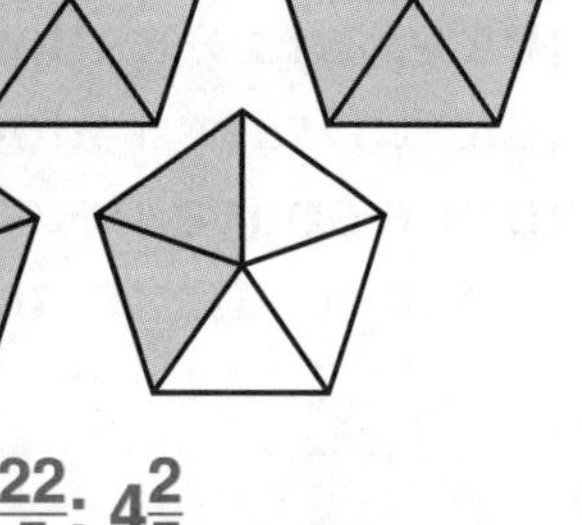

$\frac{22}{5}$; $4\frac{2}{5}$

4.

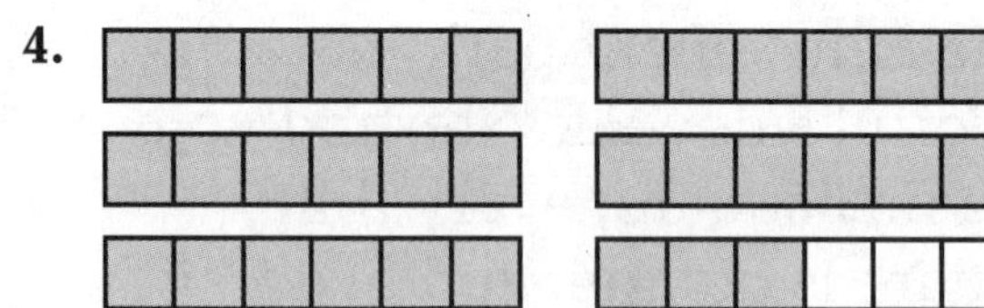

$\frac{33}{6}$; $5\frac{3}{6}$ or $5\frac{1}{2}$

Rename each fraction as a mixed number. You may wish to draw a picture.

5. $\frac{3}{2}$ $1\frac{1}{2}$

6. $\frac{13}{5}$ $2\frac{3}{5}$

7. $\frac{22}{4}$ $5\frac{2}{4}$ or $5\frac{1}{2}$

8. $\frac{11}{3}$ $3\frac{2}{3}$

9. $\frac{26}{6}$ $4\frac{2}{6}$ or $4\frac{1}{3}$

10. $\frac{18}{8}$ $2\frac{2}{8}$ or $2\frac{1}{4}$

Name ______________________

Adding Like Fractions

When you add like fractions, add only the numerators.

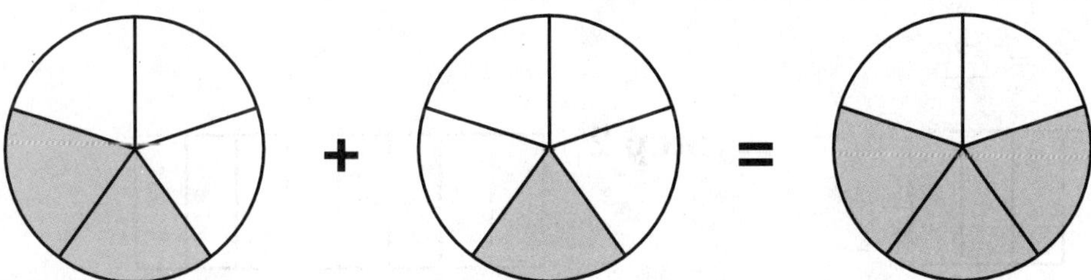

$\frac{\text{2 parts shaded}}{\text{5 parts}} + \frac{\text{1 part shaded}}{\text{5 parts}} = \frac{\text{3 parts shaded}}{\text{5 parts}}$

$\frac{2}{5} + \frac{1}{5}$ → Add the numerators. → $\frac{2+1}{5}$ → $= \frac{3}{5}$
→ Write the denominator. → →

Find the sum in Exercises 2–14. Show how you added the numerators.

1. $\frac{2}{6} + \frac{3}{6} = \frac{2+3}{6} = \frac{5}{6}$
2. $\frac{3}{7} + \frac{3}{7} = \frac{3+3}{7} = \frac{6}{7}$
3. $\frac{1}{8} + \frac{4}{8} = \frac{1+4}{8} = \frac{5}{8}$
4. $\frac{2}{5} + \frac{2}{5} = \frac{2+2}{5} = \frac{4}{5}$
5. $\frac{2}{4} + \frac{1}{4} = \frac{2+1}{4} = \frac{3}{4}$
6. $\frac{1}{3} + \frac{1}{3} = \frac{1+1}{3} = \frac{2}{3}$
7. $\frac{4}{10} + \frac{5}{10} = \frac{4+5}{10} = \frac{9}{10}$
8. $\frac{4}{9} + \frac{1}{9} = \frac{4+1}{9} = \frac{5}{9}$
9. $\frac{4}{12} + \frac{6}{12} = \frac{4+6}{12} = \frac{10}{12}$
10. $\frac{2}{8} + \frac{1}{8} = \frac{2+1}{8} = \frac{3}{8}$
11. $\frac{3}{5} + \frac{2}{5} = \frac{3+2}{5} = \frac{5}{5}$, or 1
12. $\frac{5}{8} + \frac{6}{8} = \frac{5+6}{8} = \frac{11}{8}$, or $1\frac{3}{8}$
13. $\frac{3}{9} + \frac{3}{9} = \frac{3+3}{9} = \frac{6}{9}$
14. $\frac{1}{7} + \frac{5}{7} = \frac{1+5}{7} = \frac{6}{7}$

Name ______________________________

LESSON 21.2

More About Adding Like Fractions

You can add like fractions by making a model and shading parts of it to represent the fractions in the problem.

Find the sum. $\frac{1}{8} + \frac{2}{8} = \underline{?}$

Step 1

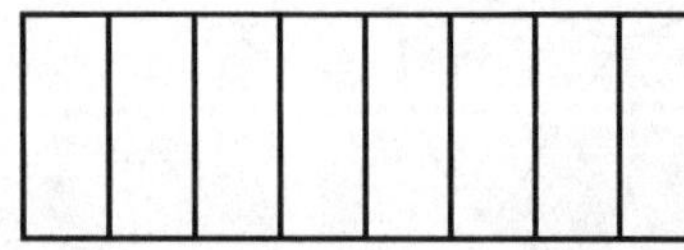

Since both fractions have an 8 in the denominator, divide a rectangle into 8 equal parts.

Step 2

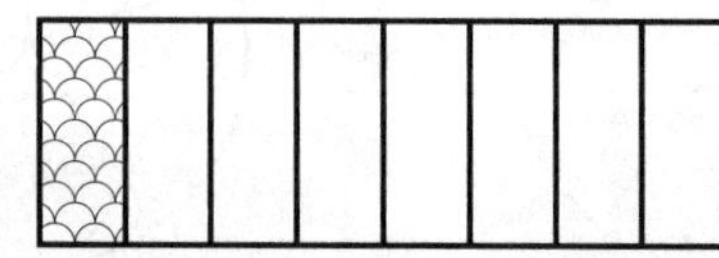

For the first fraction, $\frac{1}{8}$, shade 1 part out of 8 parts.

Step 3

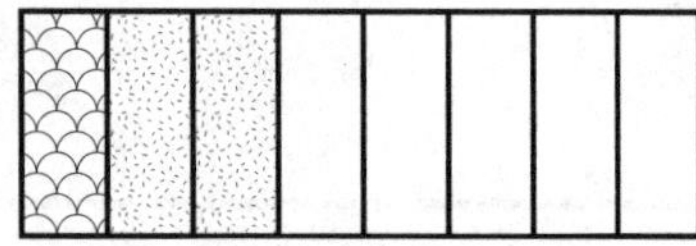

For the second fraction, $\frac{2}{8}$, shade 2 more parts out of 8 parts.

Step 4

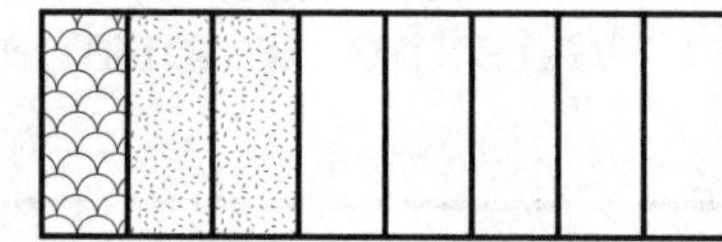

Count the number of shaded parts. There are 3 parts out of 8 parts shaded. So, $\frac{1}{8} + \frac{2}{8} = \frac{3}{8}$.

Create a drawing for each problem and find the sum. **Check students' drawings.**

1. $\frac{2}{6} + \frac{3}{6} = \underline{\frac{5}{6}}$

2. $\frac{2}{8} + \frac{4}{8} = \underline{\frac{6}{8}}$

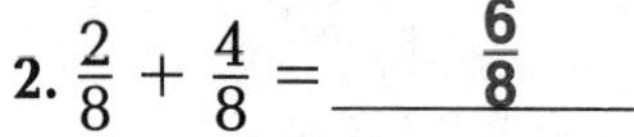

3. $\frac{2}{5} + \frac{3}{5} = \underline{\frac{5}{5}\text{, or } 1}$

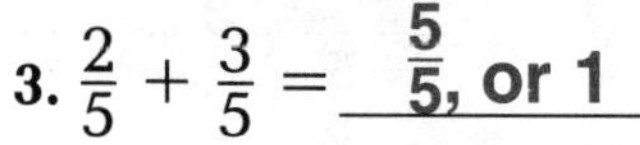

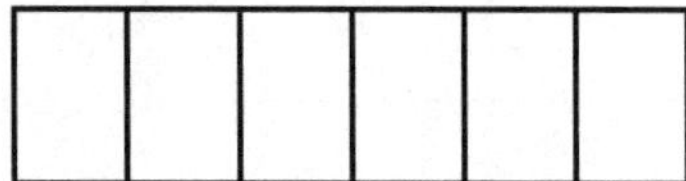

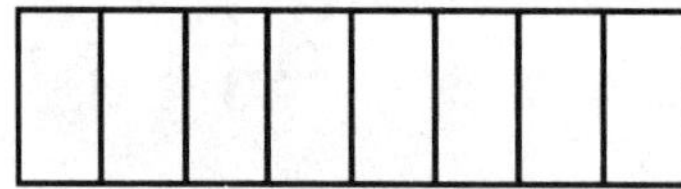

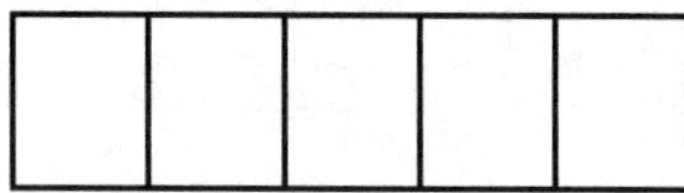

4. $\frac{1}{4} + \frac{2}{4} = \underline{\frac{3}{4}}$

5. $\frac{1}{3} + \frac{1}{3} = \underline{\frac{2}{3}}$

6. $\frac{1}{2} + \frac{1}{2} = \underline{\frac{2}{2}\text{, or } 1}$

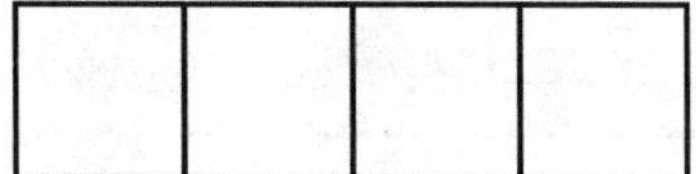

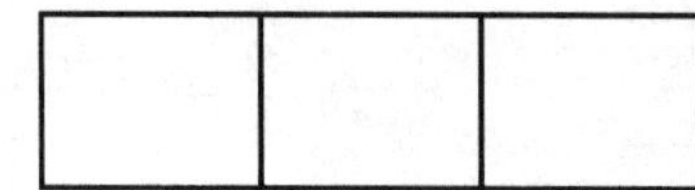

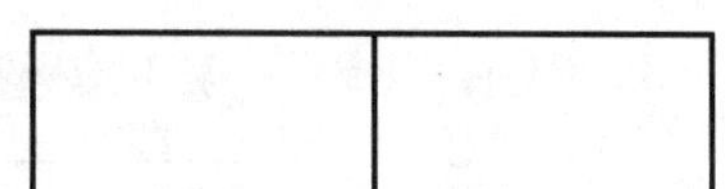

7. $\frac{3}{10} + \frac{5}{10} = \underline{\frac{8}{10}}$

8. $\frac{3}{12} + \frac{6}{12} = \underline{\frac{9}{12}}$

9. $\frac{5}{9} + \frac{2}{9} = \underline{\frac{7}{9}}$

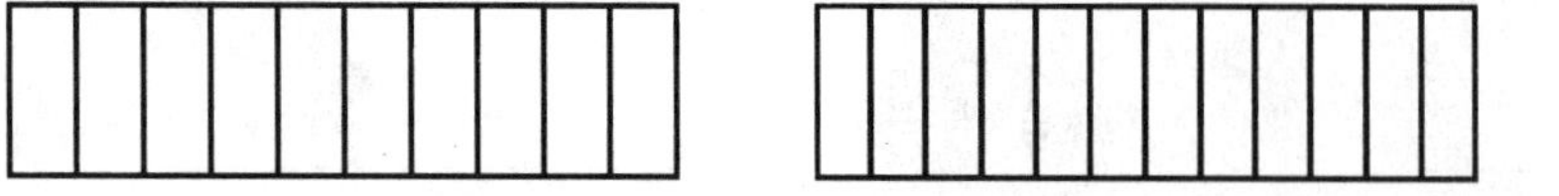

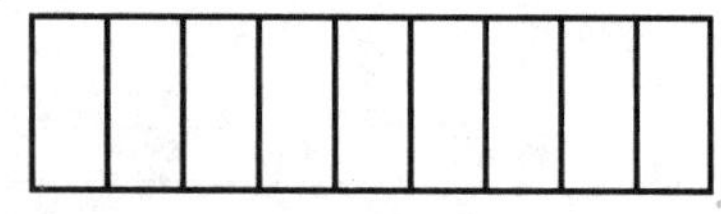

Name ______________________________

Subtracting Like Fractions

When you subtract like fractions, subtract only the numerators.

$\frac{6}{8} - \frac{2}{8} = \underline{?}$

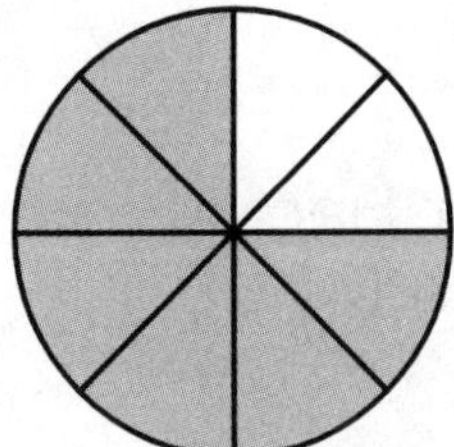

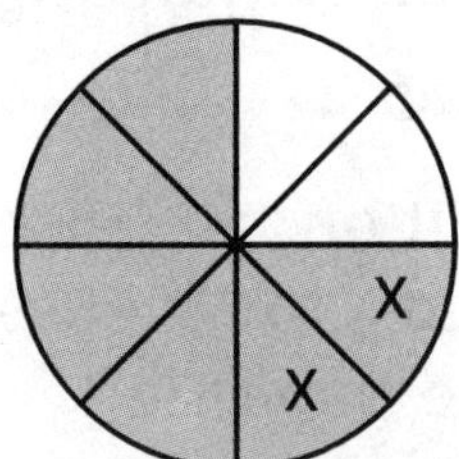

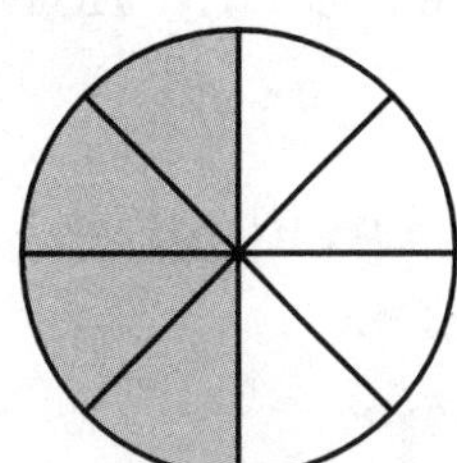

$\frac{\text{6 parts shaded}}{\text{8 parts}} - \frac{\text{2 parts shaded}}{\text{8 parts}} = \frac{\text{4 parts shaded}}{\text{8 parts}}$

$\frac{6}{8} - \frac{2}{8}$ → Subtract the numerators. → $\frac{6-2}{8}$ → $= \frac{4}{8}$
→ Write the denominator. → →

Find the difference. Show how you subtracted the numerators.

1. $\frac{5}{6} - \frac{3}{6} = \frac{5-3}{6} = \frac{2}{6}$
2. $\frac{6}{7} - \frac{3}{7} = \frac{6-3}{7} = \frac{3}{7}$
3. $\frac{7}{8} - \frac{4}{8} = \frac{7-4}{8} = \frac{3}{8}$
4. $\frac{4}{5} - \frac{2}{5} = \frac{4-2}{5} = \frac{2}{5}$
5. $\frac{2}{4} - \frac{1}{4} = \frac{2-1}{4} = \frac{1}{4}$
6. $\frac{2}{3} - \frac{1}{3} = \frac{2-1}{3} = \frac{1}{3}$
7. $\frac{7}{10} - \frac{5}{10} = \frac{7-5}{10} = \frac{2}{10}$
8. $\frac{4}{9} - \frac{1}{9} = \frac{4-1}{9} = \frac{3}{9}$
9. $\frac{10}{12} - \frac{6}{12} = \frac{10-6}{12} = \frac{4}{12}$
10. $\frac{2}{8} - \frac{1}{8} = \frac{2-1}{8} = \frac{1}{8}$
11. $\frac{9}{10} - \frac{3}{10} = \frac{9-3}{10} = \frac{6}{10}$
12. $\frac{9}{12} - \frac{1}{12} = \frac{9-1}{12} = \frac{8}{12}$
13. $\frac{7}{9} - \frac{5}{9} = \frac{7-5}{9} = \frac{2}{9}$
14. $\frac{5}{5} - \frac{4}{5} = \frac{5-4}{5} = \frac{1}{5}$
15. $\frac{3}{4} - \frac{1}{4} = \frac{3-1}{4} = \frac{2}{4}$
16. $\frac{7}{8} - \frac{2}{8} = \frac{7-2}{8} = \frac{5}{8}$

Name ____________________

Problem-Solving Strategy

Make a Model

Hector and Melissa each took $\frac{1}{8}$ of the juice in the container. What fraction of the juice was left?

You can *make a model* to find how much of the juice was left.

Step 1

Use a piece of paper. Divide the paper into 8 equal parts.

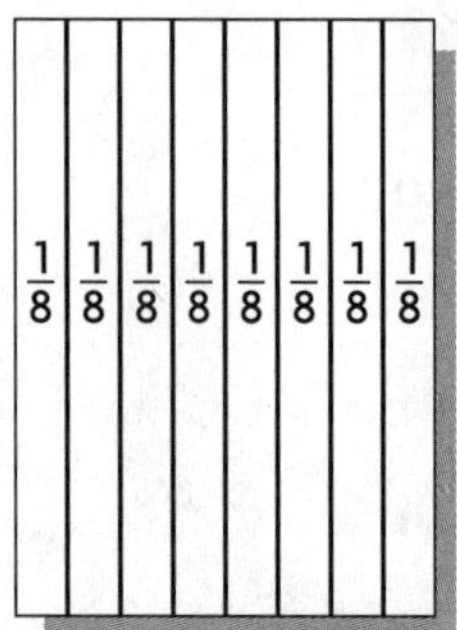

Step 2

Cut off two pieces, $\frac{1}{8}$ for Hector and $\frac{1}{8}$ for Melissa. There is $\frac{6}{8}$ of the paper left.

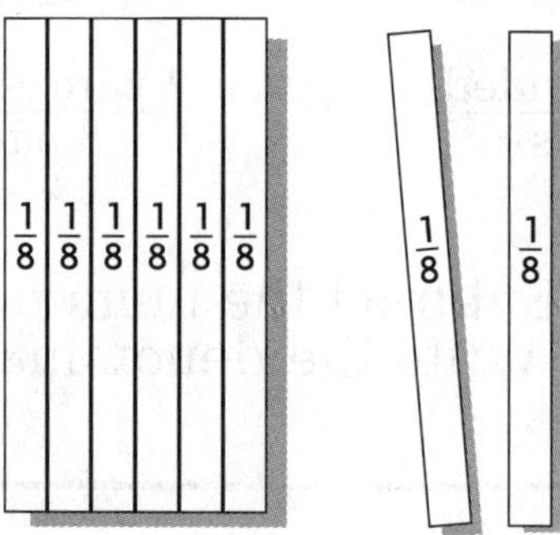

So, $\frac{6}{8}$ of the juice in the container was left.

Make a model and solve.

1. John and Peter each took $\frac{2}{5}$ of the ribbon. How much ribbon was left?

   $\frac{1}{5}$ **of the ribbon**

2. Ravi is putting up a fence 24 meters long. The fenceposts are 3 meters apart. How many fenceposts does he need?

   9 fence posts

3. Twelve friends got together to play soccer. In the group $\frac{3}{12}$ wore red shirts and $\frac{4}{12}$ wore blue shirts. The rest wore white shirts. What fraction of the friends wore white shirts?

   $\frac{5}{12}$ **wore white shirts**

4. Jim folded a strip of paper in half, then in half again, and then in half once more. When he opened up the folded strip of paper, how many sections did he see?

   8 sections

Name ____________________

LESSON 21.4

Adding Mixed Numbers

When adding mixed numbers, first add the fractions and then add the whole numbers.

Find the sum.	**Step 1**	**Step 2**
	Add the fractions.	Add the whole numbers.
$\begin{array}{r} 7\frac{2}{6} \\ +4\frac{3}{6} \\ \hline \end{array}$	$\begin{array}{r} 7\frac{2}{6} \\ +4\frac{3}{6} \\ \hline \frac{5}{6} \end{array}$	$\begin{array}{r} 7\frac{2}{6} \\ +4\frac{3}{6} \\ \hline 11\frac{5}{6} \end{array}$

So, the sum is $11\frac{5}{6}$.

If the problem is written horizontally, you can rewrite the problem vertically and then follow the above steps.

Write $3\frac{1}{7} + 4\frac{5}{7} = \underline{\ ?\ } \rightarrow \begin{array}{r} 3\frac{1}{7} \\ +4\frac{5}{7} \\ \hline \end{array} \quad \begin{array}{r} 3\frac{1}{7} \\ +4\frac{5}{7} \\ \hline 7\frac{6}{7} \end{array}$

Find the sum.

1. $\begin{array}{r} 5\frac{2}{9} \\ +2\frac{4}{9} \\ \hline 7\frac{6}{9} \end{array}$

2. $\begin{array}{r} 3\frac{1}{6} \\ +7\frac{3}{6} \\ \hline 10\frac{4}{6} \end{array}$

3. $\begin{array}{r} 5\frac{3}{10} \\ +2\frac{6}{10} \\ \hline 7\frac{9}{10} \end{array}$

4. $\begin{array}{r} 7\frac{1}{8} \\ +1\frac{3}{8} \\ \hline 8\frac{4}{8} \end{array}$

5. $\begin{array}{r} 8\frac{1}{5} \\ +3\frac{1}{5} \\ \hline 11\frac{2}{5} \end{array}$

6. $\begin{array}{r} 9\frac{3}{7} \\ +7\frac{2}{7} \\ \hline 16\frac{5}{7} \end{array}$

7. $\begin{array}{r} 4\frac{5}{9} \\ +3\frac{1}{9} \\ \hline 7\frac{6}{9} \end{array}$

8. $\begin{array}{r} 6\frac{2}{4} \\ +3\frac{1}{4} \\ \hline 9\frac{3}{4} \end{array}$

9. $5\frac{2}{8} + 6\frac{3}{8} = \underline{\ ?\ }$ $11\frac{5}{8}$

10. $8\frac{3}{12} + 2\frac{8}{12} = \underline{\ ?\ }$ $10\frac{11}{12}$

11. $7\frac{1}{4} + 3\frac{2}{4} = \underline{\ ?\ }$ $10\frac{3}{4}$

Name ______________________________

LESSON 21.5

Subtracting Mixed Numbers

When subtracting mixed numbers, first subtract the fractions and then subtract the whole numbers.

Find the difference. **Step 1** Subtract the fractions. **Step 2** Subtract the whole numbers.

$$\begin{array}{r} 8\frac{5}{6} \\ -\ 1\frac{3}{6} \\ \hline \end{array} \qquad \begin{array}{r} 8\frac{5}{6} \\ -\ 1\frac{3}{6} \\ \hline \frac{2}{6} \end{array} \qquad \begin{array}{r} 8\frac{5}{6} \\ -\ 1\frac{3}{6} \\ \hline 7\frac{2}{6} \end{array}$$

So, the difference is $7\frac{2}{6}$.

If the problem is written horizontally, you can rewrite the problem vertically and then follow the above steps.

Find the difference.

1. $\begin{array}{r} 5\frac{7}{9} \\ -2\frac{4}{9} \\ \hline \mathbf{3\frac{3}{9}} \end{array}$

2. $\begin{array}{r} 8\frac{3}{6} \\ -7\frac{1}{6} \\ \hline \mathbf{1\frac{2}{6}} \end{array}$

3. $\begin{array}{r} 5\frac{6}{10} \\ -2\frac{3}{10} \\ \hline \mathbf{3\frac{3}{10}} \end{array}$

4. $\begin{array}{r} 9\frac{7}{8} \\ -1\frac{3}{8} \\ \hline \mathbf{8\frac{4}{8}} \end{array}$

5. $\begin{array}{r} 8\frac{4}{5} \\ -3\frac{1}{5} \\ \hline \mathbf{5\frac{3}{5}} \end{array}$

6. $\begin{array}{r} 9\frac{3}{7} \\ -7\frac{2}{7} \\ \hline \mathbf{2\frac{1}{7}} \end{array}$

7. $\begin{array}{r} 4\frac{5}{9} \\ -3\frac{1}{9} \\ \hline \mathbf{1\frac{4}{9}} \end{array}$

8. $\begin{array}{r} 6\frac{2}{4} \\ -3\frac{1}{4} \\ \hline \mathbf{3\frac{1}{4}} \end{array}$

Write vertically and find the difference.

9. $6\frac{3}{8} - 5\frac{2}{8} =$? **$1\frac{1}{8}$**

10. $8\frac{10}{12} - 1\frac{4}{12} =$? **$7\frac{6}{12}$**

11. $7\frac{3}{4} - 2\frac{1}{4} =$? **$5\frac{2}{4}$**

Name ______________________________

LESSON 22.1

Relating Fractions and Decimals

You can write a fraction or a decimal to tell what part is shaded.

Model	Fraction	Decimal			Read
		O	T	H	
	4 shaded parts / 10 parts	0 .	4		four tenths
		O	T	H	
	25 shaded parts / 100 parts	0 .	2	5	twenty-five hundredths

Complete the table.

	Model	Fraction	Decimal			Read
1.		7 shaded parts / 10 parts	O 0 .	T 7	H	seven tenths
2.		12 shaded parts / 100 parts	O 0 .	T 1	H 2	twelve hundredths
3.		2 shaded parts / 10 parts	O 0 .	T 2	H	two tenths
4.		5 shaded parts / 10 parts	O 0 .	T 5	H	five tenths
5.		38 shaded parts / 100 parts	O 0 .	T 3	H 8	thirty-eight hundredths

Name ____________________

LESSON 22.2

Tenths and Hundredths

You can write a decimal for the part that is shaded.

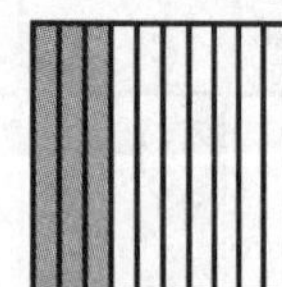	
3 out of 10 equal parts that are shaded.	41 out of 100 equal parts that are shaded.
Write: 0.3	Write: 0.41

Write the decimal for the shaded part of each model.

1. 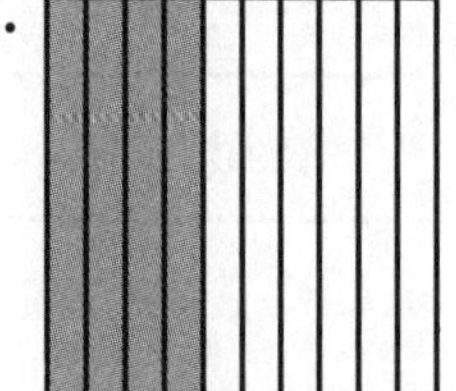  0.4

2. 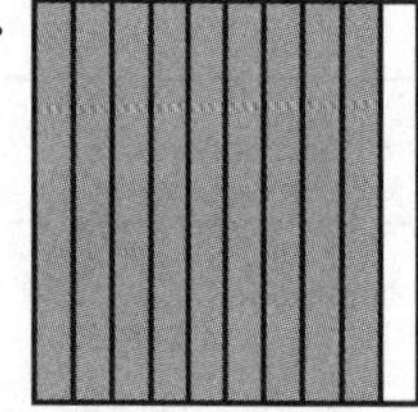 0.9

3. 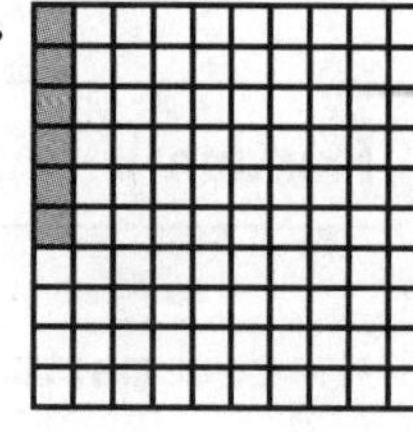 0.06

4. 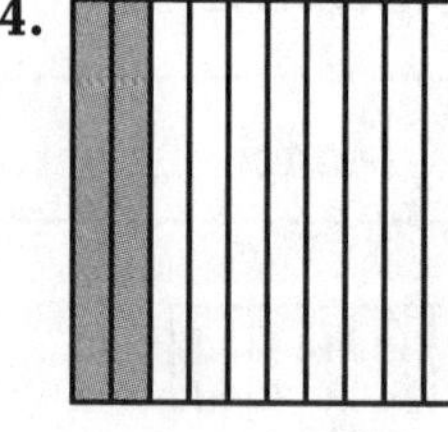 0.2

5. 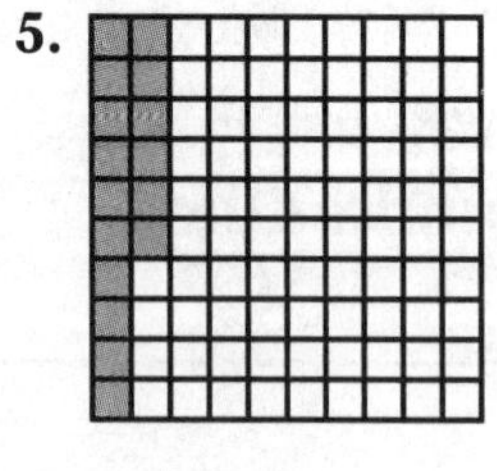 0.16

6. 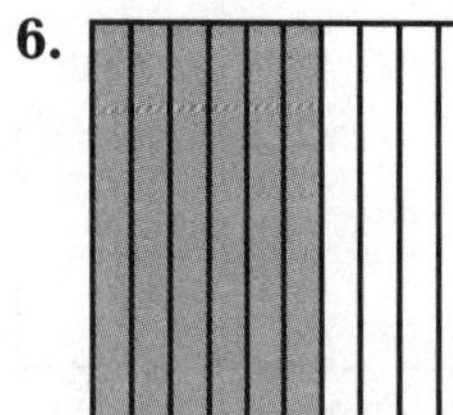 0.6

7. 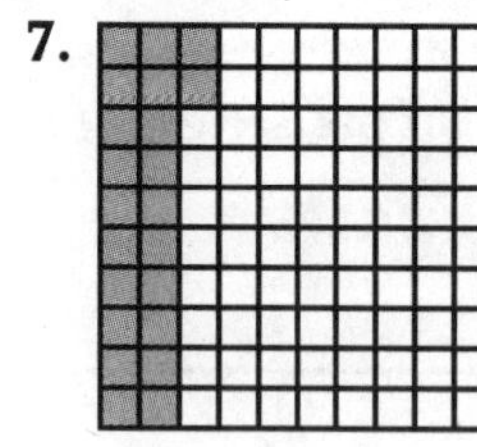 0.22

8. 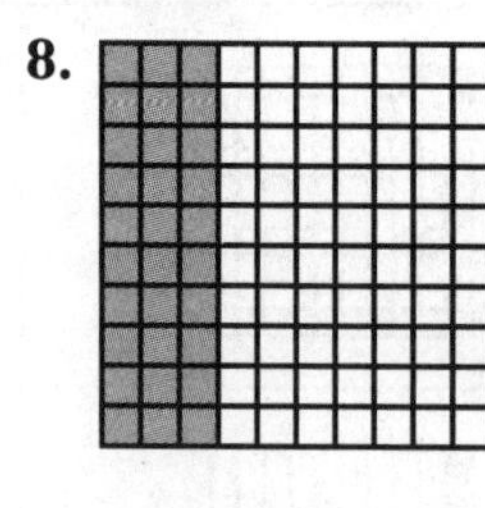 0.30

Shade each model to show the decimal amount.

9. 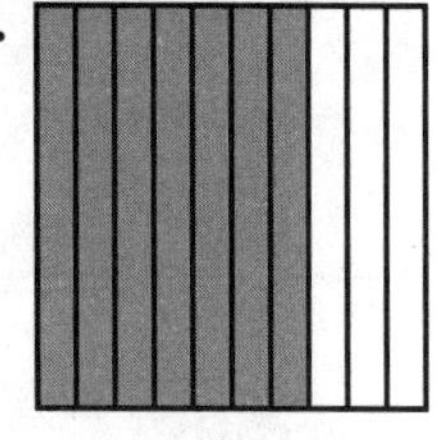 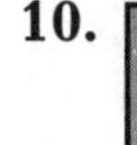 0.7

10. 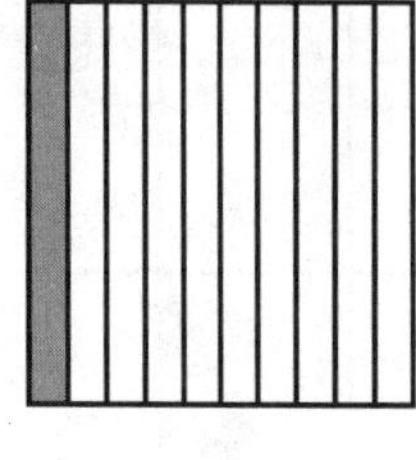 0.1

11. 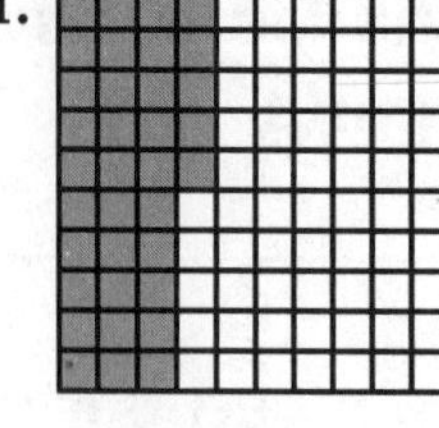 0.35

12. 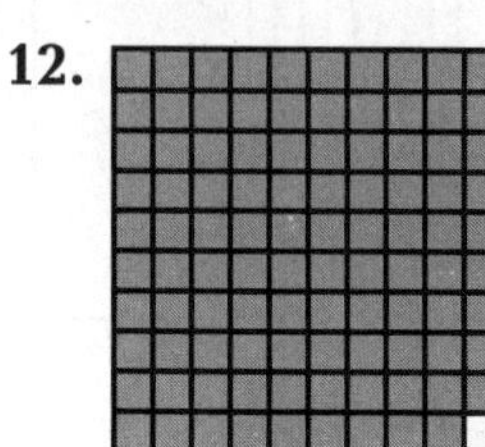 0.99

Name ______________________________

Equivalent Decimals

Equivalent decimals are different names for the same amount.

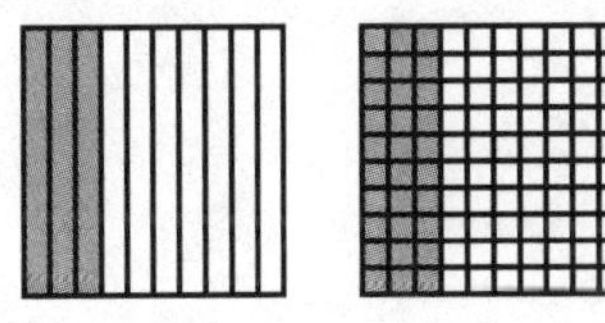

0.3 = 0.30

The decimal models show that 0.3 and 0.30 are equivalent decimals. The same amount is shaded in each model.

Ones		Tenths	Hundredths
0	.	3	
0	.	3	0

The place-value chart also shows that 0.3 = 0.30. Both numbers have the digit 3 in the tenths place.

Shade an equivalent amount in each pair of models. Write the decimal for each shaded amount.

1. 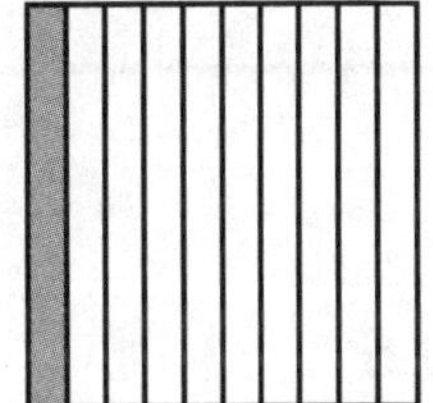

0.1 = 0.10

2. 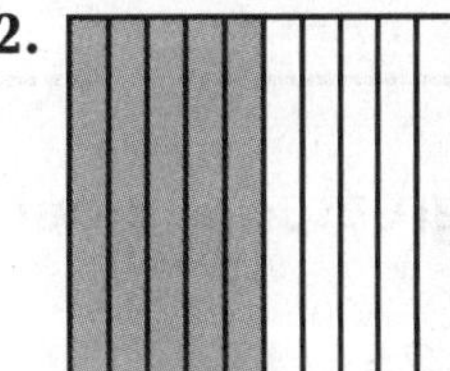

0.5 = 0.50

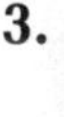

3. 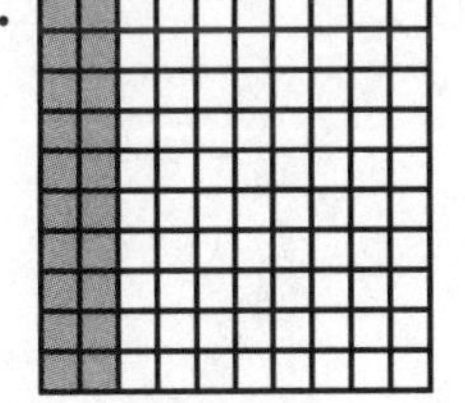

0.20 = 0.2

4. 

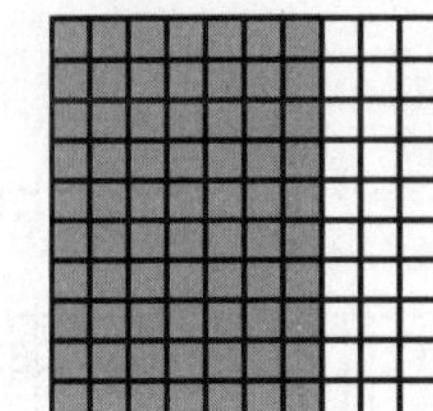

0.7 = 0.70

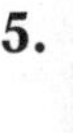

5. 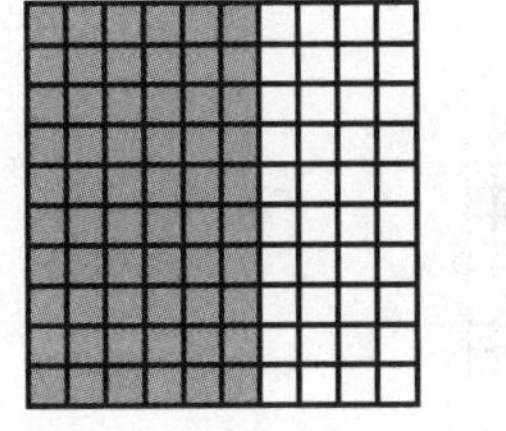

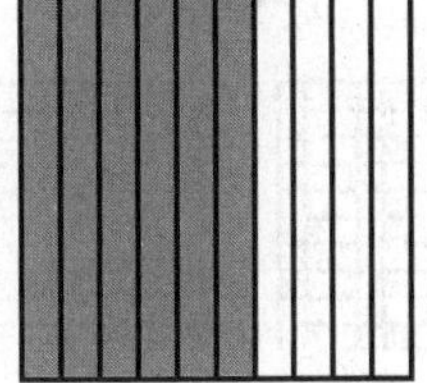

0.60 = 0.6

6. 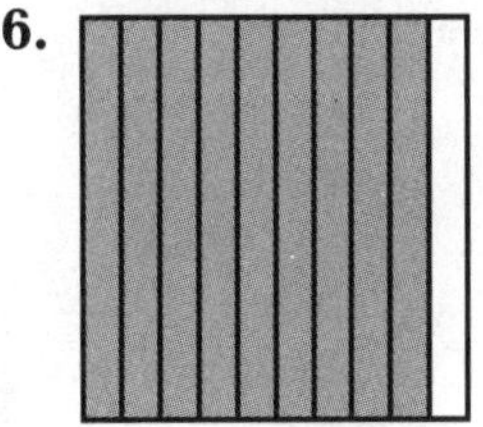

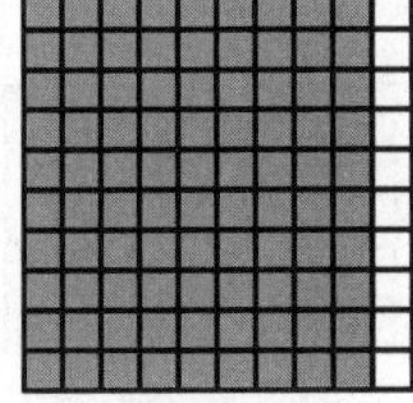

0.9 = 0.90

Name ______________________________________

Comparing and Ordering

Write these numbers in order from greatest to least: 0.4, 0.35, 0.37, 0.05.

Step 1 Write the numbers in a place-value chart.

Step 2 Write a zero in the empty hundredths place.

Ones		Tenths	Hundredths
0	.	4	**0**
0	.	3	5
0	.	3	7
0	.	0	5

Step 3 Compare the digits in each place, beginning at the left.

Greatest → 0.40, 0.37, 0.35, 0.05 ← Least

Compare. Then write the greater number.

1. 0.2, 0.3 **0.3**
2. 0.25, 0.27 **0.27**
3. 0.36, 0.34 **0.36**
4. 0.5, 0.9 **0.9**
5. 0.37, 0.3 **0.37**
6. 0.02, 0.10 **0.10**
7. 0.38, 0.40 **0.40**
8. 0.04, 0.2 **0.2**

Write $<$, $>$, or $=$ in the ◯.

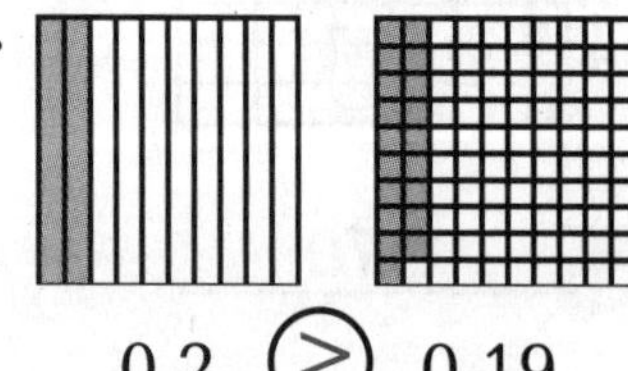

9. 0.2 $>$ 0.19

10.  0.50 $>$ 0.39

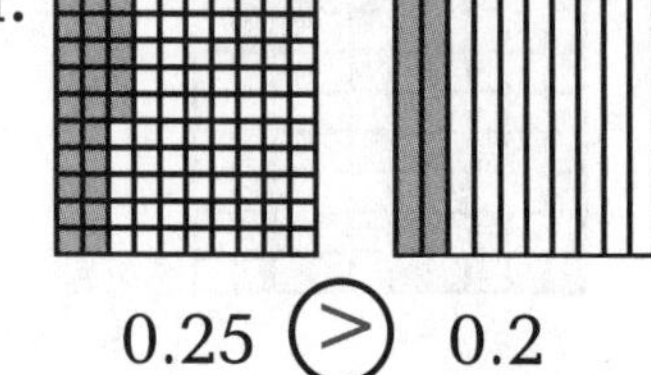

11. 0.25 $>$ 0.2

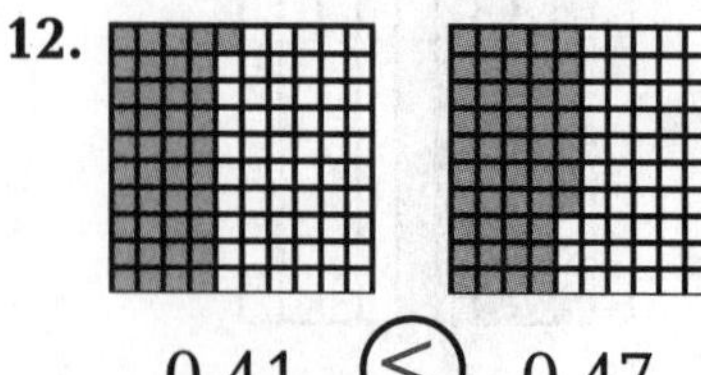

12. 0.41 $<$ 0.47

13. 0.34 $<$ 0.39
14. 0.1 $>$ 0.09
15. 0.24 $<$ 0.34
16. 0.04 $<$ 0.4

Name ______________________________

LESSON 22.5

Mixed Decimals

A **mixed number** is made up of a whole number and a fraction.

$6\frac{1}{10}$

A **mixed decimal** is a number that is made up of a whole number and a decimal.

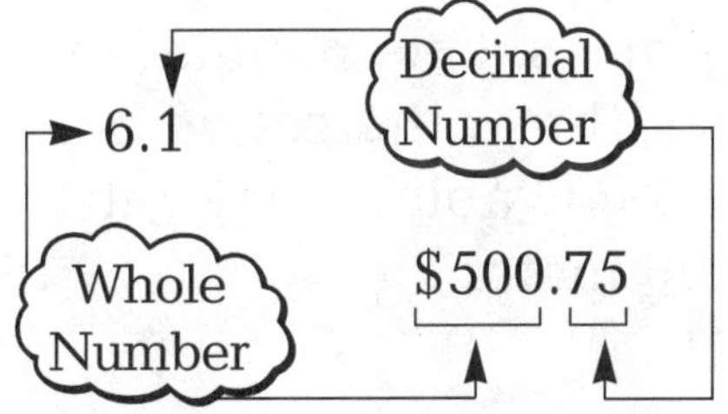

Two whole blocks and three tenths of a block are shaded.

Write: $2\frac{3}{10}$, or 2.3

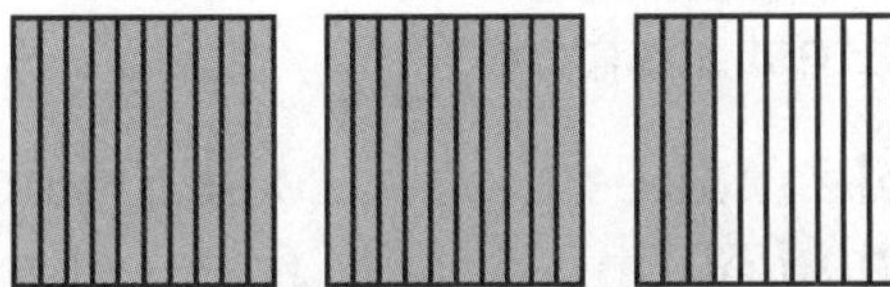

Write a mixed number and a mixed decimal for the shaded amounts.

1. 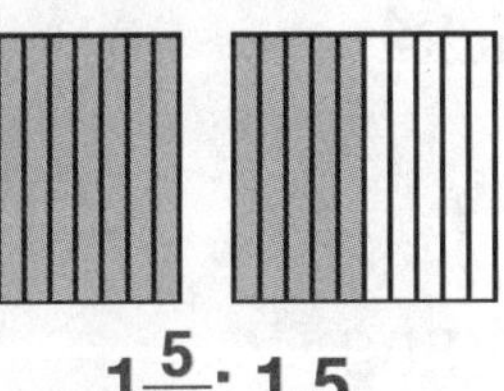

   $1\frac{5}{10}$; 1.5

2. 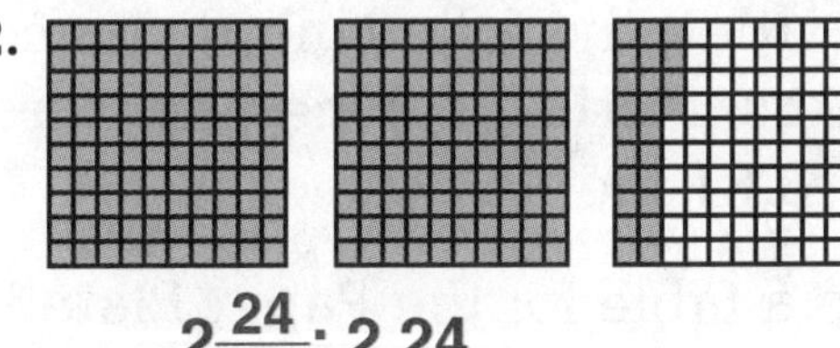

   $2\frac{24}{100}$; 2.24

3. 

   $2\frac{1}{10}$; 2.1

4. 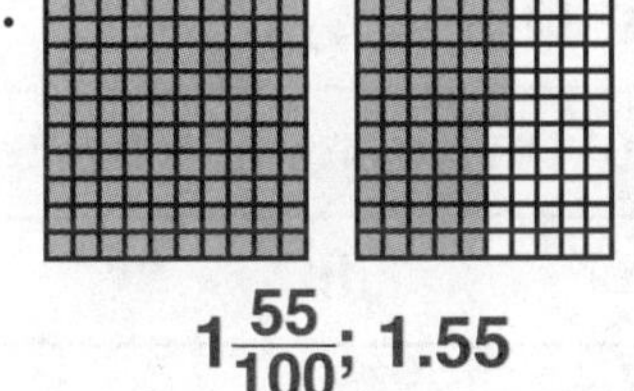

   $1\frac{55}{100}$; 1.55

Write each mixed number as a mixed decimal.

5. $3\frac{4}{10}$ **3.4**
6. $9\frac{23}{100}$ **9.23**
7. $4\frac{4}{10}$ **4.4**
8. $3\frac{9}{100}$ **3.09**

Write each mixed decimal as a mixed number.

9. 5.6 **$5\frac{6}{10}$**
10. 7.02 **$7\frac{2}{100}$**
11. 3.78 **$3\frac{78}{100}$**
12. 6.09 **$6\frac{9}{100}$**
13. 2.8 **$2\frac{8}{10}$**
14. 6.07 **$6\frac{7}{100}$**
15. 8.02 **$8\frac{2}{100}$**
16. 9.4 **$9\frac{4}{10}$**

Name ______________________

LESSON 22.5

Problem-Solving Strategy
Make a Table

The students in Ms. Turner's math class participated in a Metric Olympics. One popular event was the Cotton Ball Throw. Nick threw the cotton ball 0.91 meter. Noria threw it 1.45 meters; Caroline threw it 1.39 meters; and Graham threw it 0.98 meter.

Which student threw the cotton ball the greatest distance?

The table makes it easy to see who threw the cottonball the greatest and the least distances. So, Noria threw the cotton ball the greatest distance.

COTTON BALL THROW	
Name	**Distance in Meters**
Nick	0.91
Graham	0.98
Caroline	1.39
Noria	1.45

Another popular event was the Paper Plate Discus Throw. Nick recorded the following distances for this event. Nick–3.95 meters; Noria–2.80 meters; Caroline–4.15 meters; Graham–3.81 meters.

1. Make a table for the Paper Plate Discus Throw, arranging the distances in order from the least to the greatest.

Paper Plate Discus Throw	
Name	**Distance in Meters**
Noria	2.80
Graham	3.81
Nick	3.95
Caroline	4.15

2. Who threw the paper plate the farthest? Caroline

3. Who came in second in this event? Nick

4. How many students threw the paper plate farther than 3 meters? 3 students

Name ____________________

Modeling Addition and Subtraction

An important thing to remember when adding and subtracting decimals is to keep the place value of numbers lined up.

Find the sum. 0.82 + 0.13 Hint: Let **O** = ones, **T** = tenths, **H** = hundredths

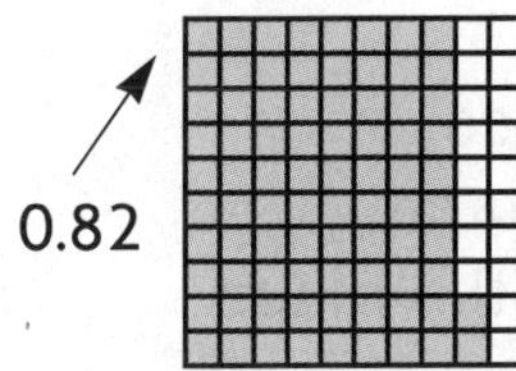

0.82

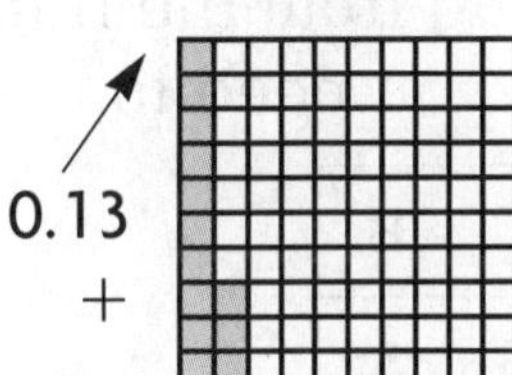

0.13
+

Step 1 Line up the decimal points.

	O	T	H
	0	. 8	2
+	0	. 1	3

Step 2 Add, starting from the right. Work your way to the left.

	O	T	H
	0	. 8	2
+	0	. 1	3
	0	. 9	5

So, 0.82 + 0.13 is 0.95. Subtraction of decimals uses the same two steps.

Use decimal squares to find the sum or difference.

1.

	O	T	H
	0	. 3	
+	0	. 2	
	0	. 5	

2.

	O	T	H
	0	. 4	6
+	0	. 3	1
	0	. 7	7

3.

	O	T	H
	0	. 3	4
+	0	. 6	6
	1	. 0	0

4.

	O	T	H
	0	. 8	
+	0	. 1	
	0	. 9	

5.

	O	T	H
	0	. 9	
−	0	. 2	
	0	. 7	

6.

	O	T	H
	0	. 8	7
−	0	. 3	6
	0	. 5	1

7.

	O	T	H
	0	. 7	
−	0	. 5	
	0	. 2	

8.

	O	T	H
	0	. 8	0
−	0	. 7	2
	0	. 0	8

Name ______________________________

LESSON 23.2

Adding Decimals

How can you add 1.65 to 0.37?

Remember to keep numbers lined up by place value and their decimal points.

Step 1 Identify the place value of each number.

1.65	0.37
1 is in the ones place.	0 is in the ones place.
6 is in the tenths place.	3 is in the tenths place.
5 is in the hundredths place.	7 is in the hundredths place.

Step 2 Line up each number by its place value.

	O	T	H
	1 .	6	5
+	0 .	3	7
	.		

Step 3 Add, starting from the right. Regroup if you need to.

	O	T	H
	[1]1 .	[1]6	5
+	0 .	3	7
	2 .	0	2

So, 1.65 + 0.37 equals 2.02.

Rewrite each problem in the place-value grid. Find the sum.

1. $0.8 + 0.5 = n$

	O	T	H
	0 .	8	
+	0 .	5	
	1 .	3	

2. $0.9 + 0.6 = n$

	O	T	H
	0 .	9	
+	0 .	6	
	1 .	5	

3. $1.67 + 0.58 = n$

	O	T	H
	1 .	6	7
+	0 .	5	8
	2 .	2	5

4. $1.46 + 0.17 = n$

	O	T	H
	1 .	4	6
+	0 .	1	7
	1 .	6	3

5. $0.6 + 1.2 = n$

	O	T	H
	0 .	6	
+	1 .	2	
	1 .	8	

6. $1.4 + 1.5 = n$

	O	T	H
	1 .	4	
+	1 .	5	
	2 .	9	

7. $1.34 + 0.82 = n$

	O	T	H
	1 .	3	4
+	0 .	8	2
	2 .	1	6

8. $1.09 + 0.8 = n$

	O	T	H
	1 .	0	9
+	0 .	8	0
	1 .	8	9

Name ___________________________

Subtracting Decimals

How can you subtract 0.58 from 1.72?

Remember to keep numbers lined up by place value and decimal points.

Step 1 Identify the place value of each number.

1.72	0.58
1 is in the ones place.	0 is in the ones place.
7 is in the tenths place.	5 is in the tenths place.
2 is in the hundredths place.	8 is in the hundredths place.

Step 2 Line up each number by its place value.

	O	T	H
	1.	7	2
−	0.	5	8
	.		

Step 3 Subtract, starting from the right. Regroup if you need to.

	O	T	H
	1.	$\overset{6}{\not 7}$	$\overset{12}{\not 2}$
−	0.	5	8
	1.	1	4

So, 1.72 − 0.58 equals 1.14.

Rewrite each problem in the place-value grid. Find the difference.

1. $0.7 - 0.1 = n$

	O	T	H
	0.	7	
−	0.	1	
	0.	6	

2. $0.9 - 0.5 = n$

	O	T	H
	0.	9	
−	0.	5	
	0.	4	

3. $1.72 - 0.51 = n$

	O	T	H
	1.	7	2
−	0.	5	1
	1.	2	1

4. $1.91 - 0.85 = n$

	O	T	H
	1.	9	1
−	0.	8	5
	1.	0	6

5. $1.7 - 1.5 = n$

	O	T	H
	1.	7	
−	1.	5	
	0.	2	

6. $1.8 - 0.9 = n$

	O	T	H
	1.	8	
−	0.	9	
	0.	9	

7. $1.29 - 0.64 = n$

	O	T	H
	1.	2	9
−	0.	6	4
	0.	6	5

8. $1.13 - 0.88 = n$

	O	T	H
	1.	1	3
−	0.	8	8
	0.	2	5

Name ____________________

LESSON 23.4

Using Decimals

How can you add 0.3 and 0.45?

Step 1 Write the decimals, keeping the decimal points in a line.

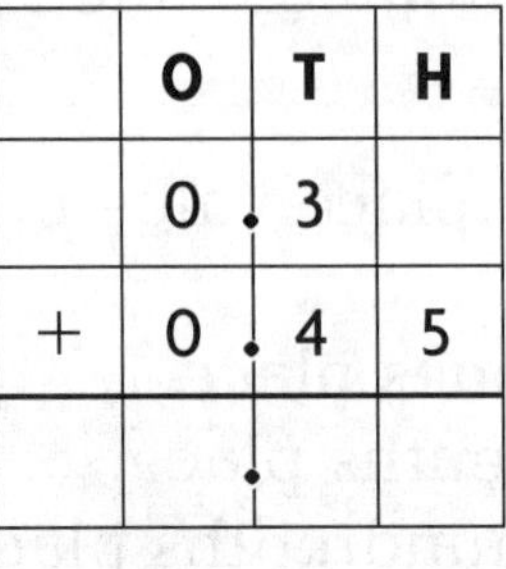

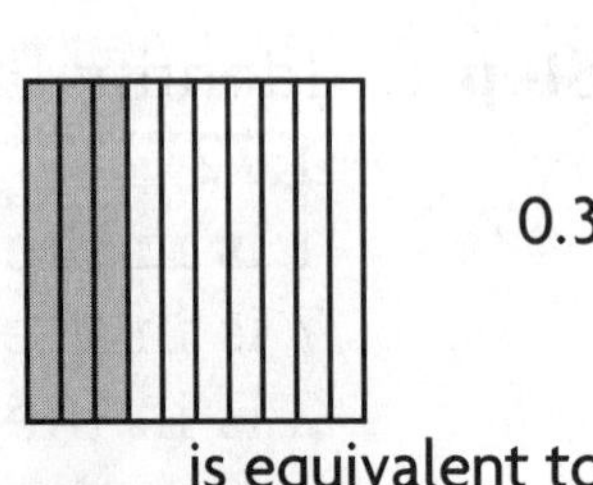

0.3

is equivalent to

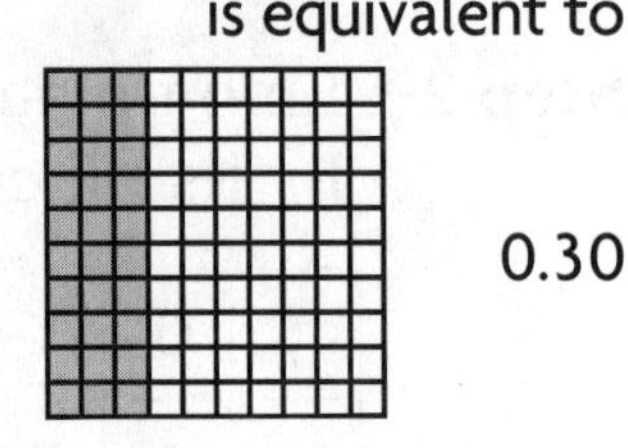

0.30

Step 2 The number 0.3 has fewer digits. Write an equivalent decimal with a zero in the hundredths place so that both numbers have the same number of digits.

	O	T	H
	0.	3	0
+	0.	4	5
	0.	7	5

Step 3 Add as you would with whole numbers.

Follow the same steps when you subtract.

Rewrite each problem in the place-value grid. Solve.

1. $1.34 + 0.5 = n$

	O	T	H
	1.	3	4
+	0.	5	0
	1.	8	4

2. $1.67 + 0.2 = n$

	O	T	H
	1.	6	7
+	0.	2	0
	1.	8	7

3. $1.2 + 1.32 = n$

	O	T	H
	1.	2	0
+	1.	3	2
	2.	5	2

4. $1.02 + 0.87 = n$

	O	T	H
	1.	0	2
+	0.	8	7
	1.	8	9

5. $1.8 - 0.52 = n$

	O	T	H
	1.	8	0
−	0.	5	2
	1.	2	8

6. $2.4 - 1.57 = n$

	O	T	H
	2.	4	0
−	1.	5	7
	0.	8	3

7. $1.20 - 0.56 = n$

	O	T	H
	1.	2	0
−	0.	5	6
	0.	6	4

8. $1.03 - 0.8 = n$

	O	T	H
	1.	0	3
−	0.	8	0
	0.	2	3

Name ______________________________

Problem-Solving Strategy

Write a Number Sentence

Carrie wants to walk to her friend Sue's house, pick her up, and then walk to Joan's house. Then all three girls will walk back to Carrie's house. It is 2.3 miles from Carrie's house to Sue's house. It is 1.6 miles from Sue's house to Joan's house and 3.1 miles from Joan's house to Carrie's house. How many miles will Carrie walk?

Write a number sentence that shows the facts of the problem.

Distance from Carrie's to Sue's house		Distance from Sue's to Joan's house		Distance from Joan's to Carrie's house	Total miles walked
2.3	+	1.6	+	3.1	$= n$

$$\begin{array}{r} 2.3 \\ 1.6 \\ +\ 3.1 \\ \hline 7.0 \end{array}$$

So, Carrie walked 7.0 miles.

For Problems 1 and 2, use the information above.
Write a number sentence to solve. **Possible number sentences are given.**

1. Sue decided to walk home from Carrie's house. What's the total distance she walked today?

   1.6 + 3.1 + 2.3 = 7.0; 7.0 mi

2. Joan decided to walk home from Carrie's house. How far did she walk today?

   3.1 + 3.1 = 6.2; 6.2 mi

Write a number sentence for each problem.
Solve, using problem-solving strategies. **Possible number sentences are given.**

3. A mystery number is added to 8 and then the answer is multiplied by 3. The product is 36. What is the mystery number?

   (8 + *n*) × 3 = 36; 4

4. Rolanda bought 8 pens for $0.65 each and a notebook for $2.76. How much money did she get back from a ten-dollar bill?

   10.00 − (5.20 + 2.76) = 2.04; $2.04

Name ______________________

LESSON 23.5

Estimating Sums and Differences by Rounding

You can estimate \$2.53 + \$1.26 + \$0.85 by rounding.

One way to estimate a sum is to round to the ones place and then add.

Step 1 Underline the digits in the ones place.
\$2.53 \$1.26 \$0.85

Step 2 Look at the digit to the right. If the digit is 5 or greater, round to the next higher whole number.
\$2.53 \$1.26 \$0.85
Round 2 up to 3. Keep as 1. Round 0 up to 1.

Step 3 Rewrite with zeros in all places to the right.
\$3.00 + \$1.00 + \$1.00 = \$5.00

Another way to estimate a sum is to round to the tenths place and then add.

\$2.53		\$1.26		\$0.85
Keep as 5.		Round 2 up to 3.		Round 8 up to 9.
\$2.50	+	\$1.30	+	\$0.90 = \$4.70

Your estimate is either \$4.70 or \$5.00, depending on how you round.

Estimate each problem twice. Then solve.

	Estimate to Ones Place	Estimate to Tenths Place	Exact Answer
1.	2.00 + 2.00 = **4.00**	2.30 + 1.60 = **3.90**	2.34 + 1.61 = **3.95**
2.	4.00 + 1.00 = **5.00**	3.50 + 1.20 = **4.70**	3.54 + 1.17 = **4.71**
3.	3.00 − 2.00 = **1.00**	2.50 − 1.70 = **0.80**	2.53 − 1.67 = **0.86**

Name ______________________

Linear Measures

When you measure length, you need to first decide which unit to use. Inches, feet, yards, and miles are all linear units which can be used to measure length.

Follow these steps to measure to the nearest inch.

Step 1

Line up one end of the object with the zero mark or the left edge of the ruler.

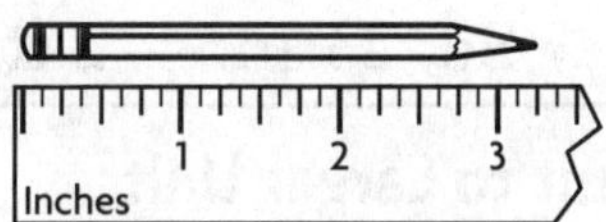

Step 2

Match the other end of the object with the number to which it is closest.

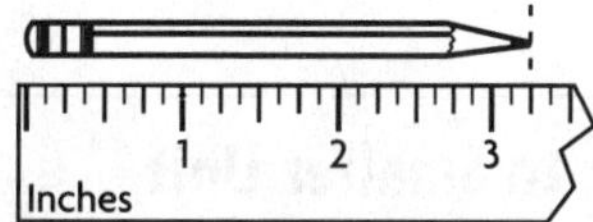

The pencil is between 3 and 4 inches long. It is closer to 3 inches long. So, to the nearest inch, the pencil is 3 inches long.

Write the measurement to the nearest inch.

1. 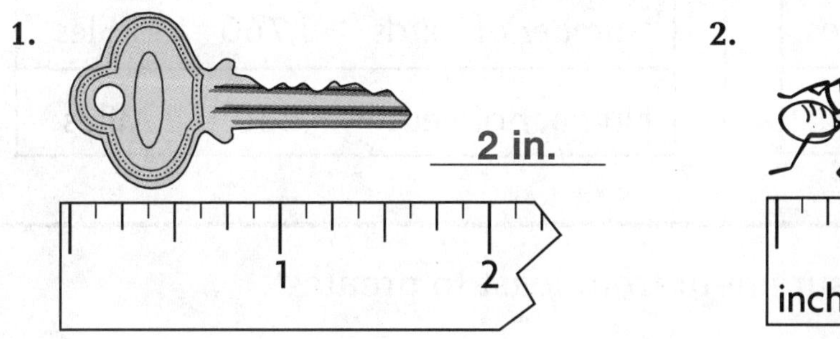

2 in.

2. 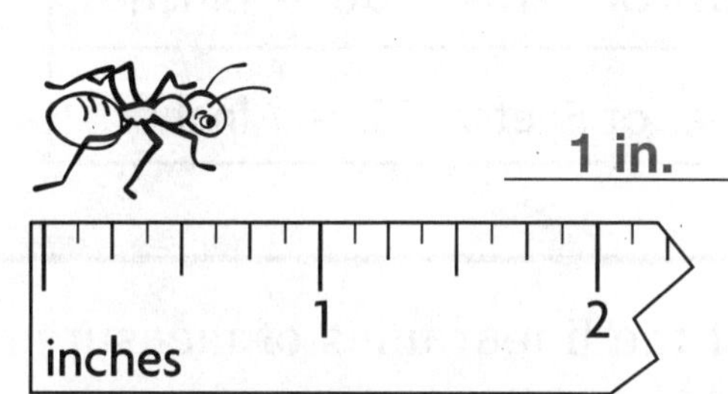

1 in.

Measure the length to the nearest inch.

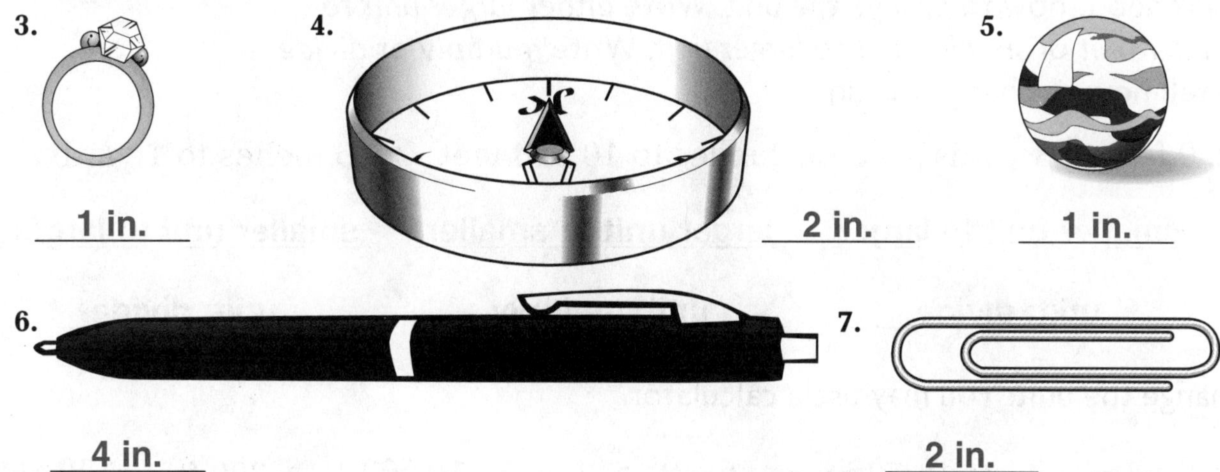

3. 1 in.

4. 2 in.

5. 1 in.

6. 4 in.

7. 2 in.

Name ______________________

Changing Units

To change units, think which measurement is larger. Then think about how the two units relate.

1 foot
1 yard

2 yd = ? ft

Yards are larger than feet. There are 3 feet in every yard. Multiply.
2 × 3 = 6 So, 2 yd = 6 ft.

1 inch
1 foot

24 in. = ? ft

Inches are smaller than feet. One foot is 12 inches. Divide.

24 ÷ 12 = 2 So, 24 in. = 2 ft.

Larger Unit to Smaller Unit
Number of Miles × 1,760 = ? Yards
Number of Miles × 5,280 = ? Feet
Number of Yards × 3 = ? Feet
Number of Yards × 36 = ? Inches
Number of Feet × 12 = ? Inches

Smaller Unit to Larger Unit
Number of Inches ÷ 12 = ? Feet
Number of Inches ÷ 36 = ? Yards
Number of Feet ÷ 3 = ? Yards
Number of Yards ÷ 1,760 = ? Miles
Number of Feet ÷ 5,280 = ? Miles

---

1. List the linear units of measurement from least to greatest.

in., ft, yd, mi

Think about how to change the unit. Write either *larger unit to smaller unit* or *smaller unit to larger unit*. Write *multiply* or *divide* to tell how to change the units.

2. 9 feet to 3 yards
smaller unit to larger unit; divide

3. 2 miles to 10,560 feet
larger unit to smaller unit; multiply

4. 36 inches to 1 yard
smaller unit to larger unit; divide

Change the unit. You may use a calculator.

5. 3 yd = 9 ft 6. 72 in. = 6 ft 7. 2 mi = 10,560 ft 8. 300 ft = 100 yd

Name ___

Problem-Solving Strategy

Draw a Diagram

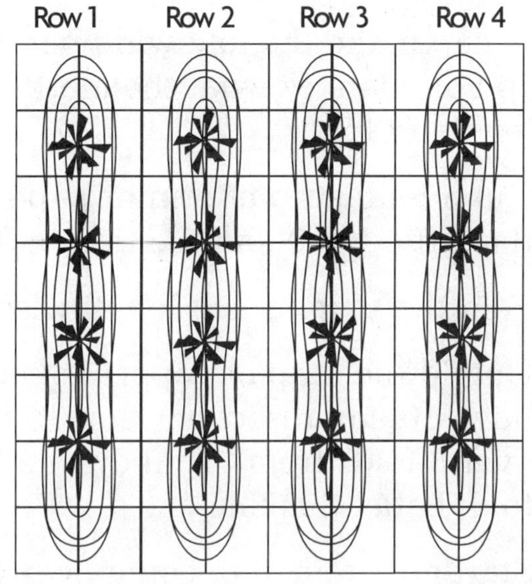

Megan is planting rows of tomato plants. The plants have to be spaced 2 feet apart. Her garden is 8 feet by 8 feet. She will start her first row 1 foot in from the edge of the garden. How many rows of tomato plants can she plant?

- To solve, use graph paper to draw a diagram of the garden. Have the length of a square represent 1 foot.
- Since each square represents 1 foot, start the first row 1 space from the edge. Draw the next row 2 feet or 2 spaces over. Continue drawing rows until no more can fit.

So, Megan can plant 4 rows.

Draw a diagram and solve.

1. Megan's tomato plants need to be spaced 2 feet apart from every other tomato plant. She will plant the first plant in each row 1 foot from the edge of the garden. How many tomato plants can Megan fit in her 8 foot by 8 foot garden?

   16 plants

2. Tom wants to plant a tomato garden like Megan's. He has a square area that is 12 feet by 12 feet. How many tomato plants can he plant?

   36 plants

3. If Megan and Tom plant marigold plants every 1 foot around the perimeter of their gardens starting at a corner, how many marigold plants will they plant?

   80 plants total; Megan = 32 plants, Tom = 48 plants

Name ______________________________

LESSON 24.3

Fractions in Measurements

YOU WILL NEED scissors, paper, ruler

You can make the following paper model to help you identify what the marks on the ruler stand for.

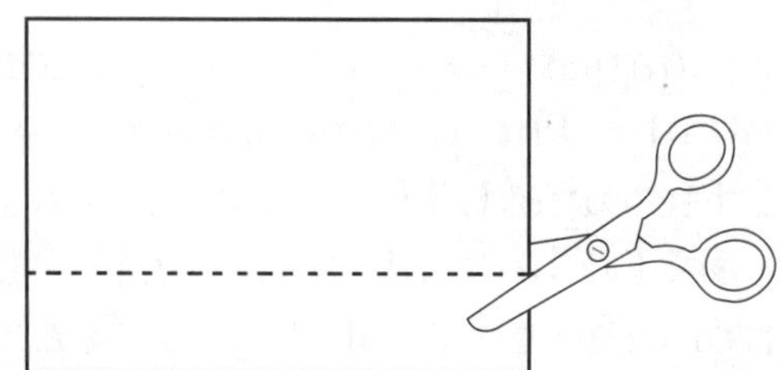

- Cut a long strip of printer paper. Label the left edge 0 and the right edge 1.

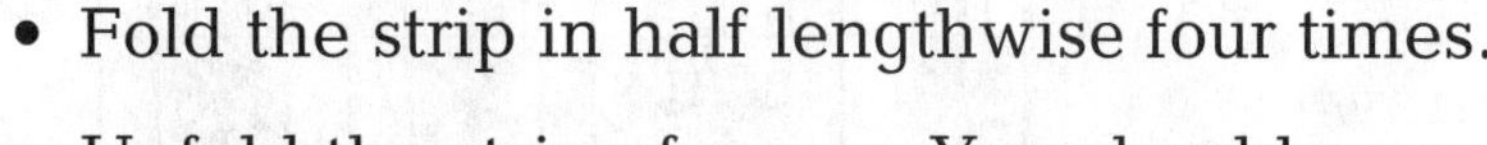

- Fold the strip in half lengthwise four times.

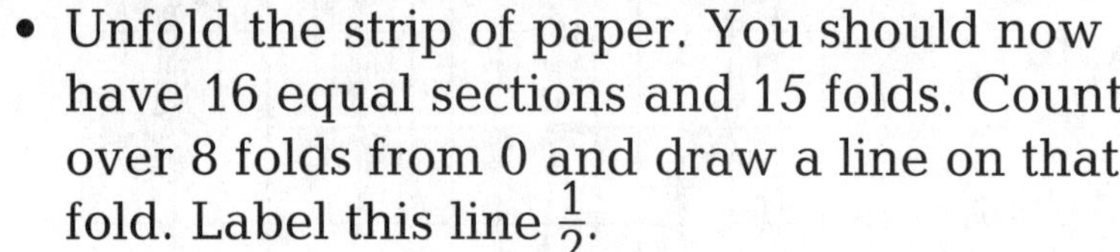

- Unfold the strip of paper. You should now have 16 equal sections and 15 folds. Count over 8 folds from 0 and draw a line on that fold. Label this line $\frac{1}{2}$.

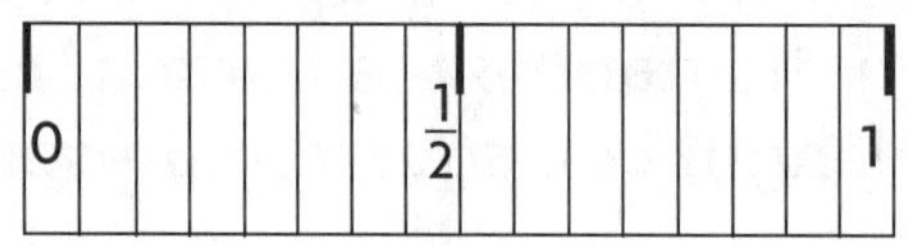

- Starting again at 0, count over 4 folds and 12 folds. Draw lines on the folds. Label the first one $\frac{1}{4}$ and the second one $\frac{3}{4}$.

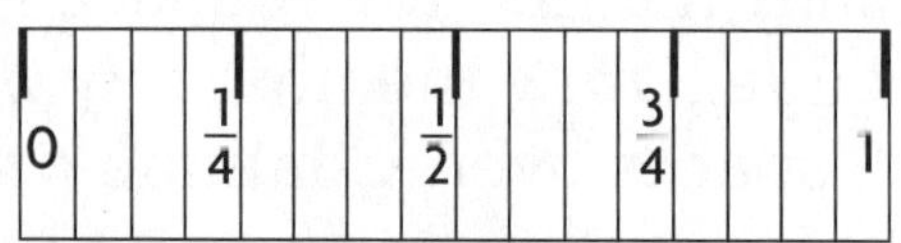

Use a ruler to measure the length to the nearest $\frac{1}{2}$ inch and to the nearest $\frac{1}{4}$ inch.

1.

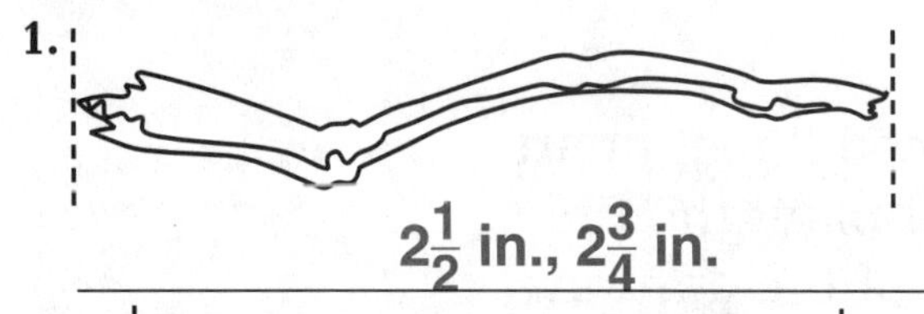

$2\frac{1}{2}$ in., $2\frac{3}{4}$ in.

2.

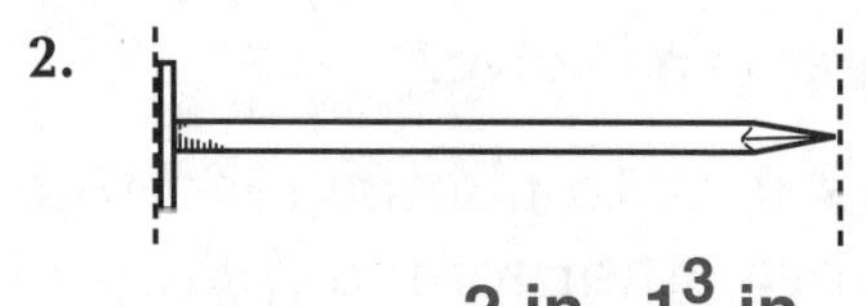

2 in., $1\frac{3}{4}$ in.

3.

$2\frac{1}{2}$ in., $2\frac{1}{4}$ in.

4.

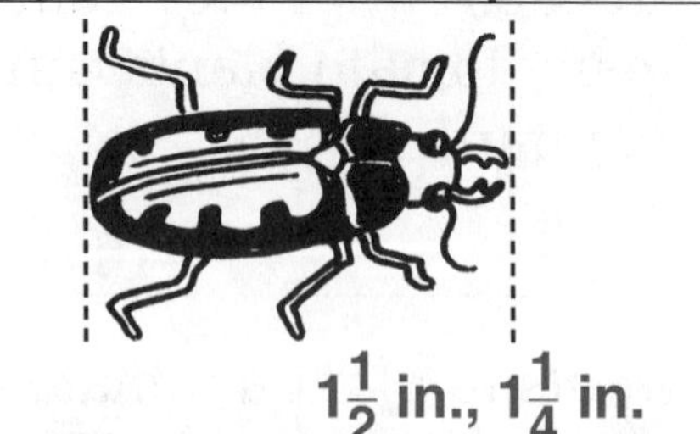

$1\frac{1}{2}$ in., $1\frac{1}{4}$ in.

5. Draw a chain of 3 large paper clips. Make each $1\frac{3}{4}$ inches long.

6. What is the length of your paper clip chain? **about $5\frac{1}{4}$ in.**

Name ________________________________

Capacity

To change units, think which unit is larger. Then think about how the two units relate.

Smaller Unit ⟶ **Larger Unit**

2 cups = 1 pint	2 pints = 1 quart	4 quarts = 1 gallon

When changing from larger units to smaller units, multiply.

2 gal = ? qt

$2 \times 4 = 8$
So, 2 gallons equal 8 quarts.

When changing from smaller units to larger units, divide.

6 c = ? pt

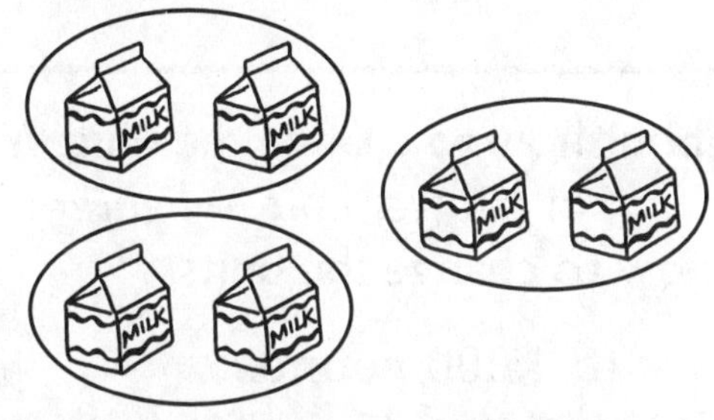

$6 \div 2 = 3$
So, 6 cups equal 3 pints.

Find the equivalent measurement by drawing a picture. **Check students' drawings.**

1. 12 c = **6** pt	2. 4 qt = **8** pt	3. 2 gal = **32** c

Name ______________________________

LESSON 24.5

Weight

Ounce	Pound	Ton
	= 16 oz	= 2,000 lb

Lightest Unit ⟶ Heaviest Unit

To change units, think which unit is heavier. Then think about how the two units relate.

3 lb = ? oz

Ounces are lighter than pounds. One pound is 16 ounces. Multiply by 16.

$3 \times 16 = 48$

So, 3 pounds equal 48 ounces.

32 oz = ? lb

Pounds are heavier than ounces. There are 16 ounces in a pound. Divide by 16.

$32 \div 16 = 2$

So, 32 ounces equal 2 pounds.

Think about how to change the unit. Write either *lighter unit to heavier unit* or *heavier unit to lighter unit*. Write *multiply* or *divide* to tell how to change the units.

1. 2 tons to 4,000 pounds
   heavier unit to lighter unit; multiply
2. 64 ounces to 4 pounds
   lighter unit to heavier unit; divide
3. 6,000 pounds to 3 tons
   lighter unit to heavier unit; divide
4. 1 pound to 16 ounces
   heavier unit to lighter unit; multiply

Circle the more reasonable measurement.

5. 

300 T or (300 lb)

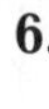
6. 

(4 oz) or 4 lb

7. 

12 oz or (12 lb)

Write the equivalent measurement. You may use a calculator.

8. 3 lb = **48** oz
9. 7 T = **14,000** lb
10. 50,000 lb = **25** T

Name ______________________

Linear Measures

The centimeter, decimeter, and meter are units of length.

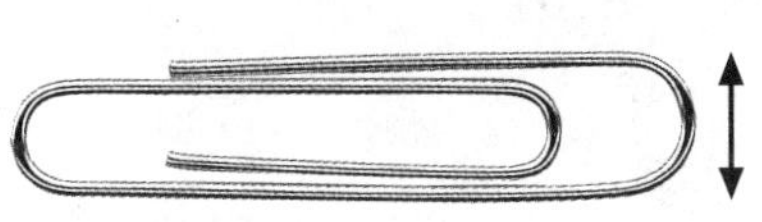

1 cm

The width of a paper clip is about 1 centimeter (cm).

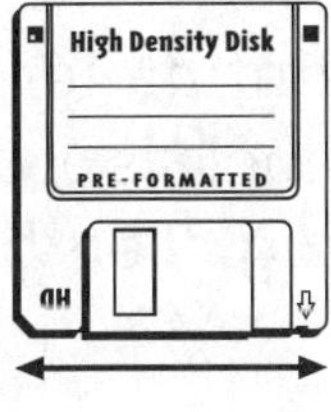

1 dm

The width of a computer disk is about 1 decimeter (dm).

1 m

The width of a door is about 1 meter (m).

Use the references above to help you decide if the measurement is close to a *cm, dm,* or *m*.

1. the thickness of a calculator

   cm

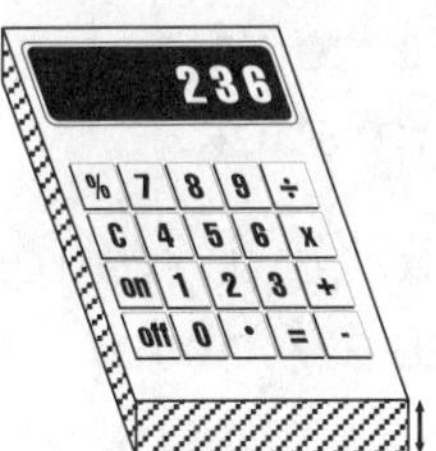

2. the width of a small paperback book dm

3. the height of a counter m

4. the length of a large dog m

5. the thickness of a workbook

   cm

6. the height of a big apple dm

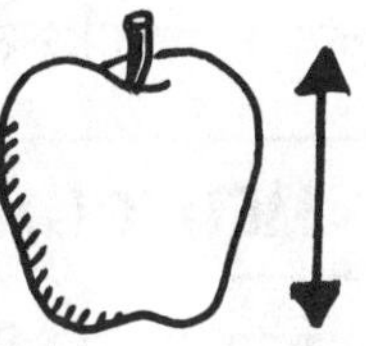

Choose the most reasonable measurement. Write *a, b,* or *c*.

7. width of a softball **a.** 1 cm **b.** 1 dm **c.** 1 m b
8. width of a basketball **a.** 25 cm **b.** 25 dm **c.** 25 m a
9. height of an average ceiling **a.** 3 cm **b.** 3 dm **c.** 3 m c
10. length of a bed **a.** 200 cm **b.** 200 dm **c.** 200 m a

Name ______________________________

LESSON 25.2

Decimals and Metric Measures

YOU WILL NEED base-ten blocks, meterstick

The base-ten blocks show how centimeters, decimeters, and meters are related. Each single block is 1 centimeter long. Each tens block is 1 decimeter long. Ten blocks lined up end to end equal 1 meter.

1 centimeter

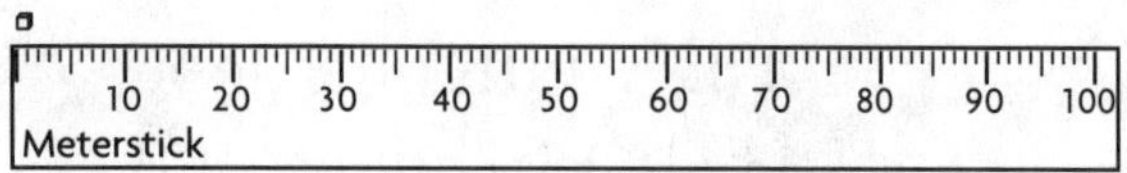

$\frac{1\text{ cm}}{100\text{ cm}} = \frac{1}{100}$ meter $= 0.01$ meter

1 decimeter = 10 centimeters = 1 tens block

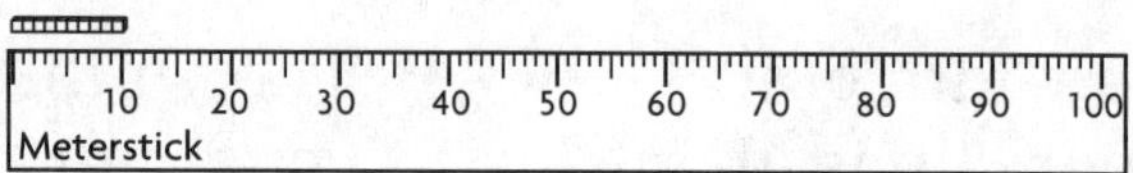

$\frac{1\text{ dm}}{10\text{ dm}} = \frac{1}{10}$ meter $= 0.1$ meter

Use blocks to help write one measurement as another.

70 cm = _?_ dm = _?_ m

70 cm = 7 dm = 0.7 m

70 cm = 0.7 m

Step 1 Line the centimeter blocks beside the meterstick.

Step 2 Count to find groups of 10 centimeters. There are 7 groups of 10 cm, so there are 7 decimeters.

Step 3 Look at the meterstick. Write the number of centimeters as a numerator over 100.

$\frac{70}{100}$ can be written as 0.70 or 0.7 meter.

Use base-ten blocks and a meterstick to complete the tables.

METRIC LINEAR UNITS			
	Centimeter	Decimeter	Meter
1.	100	10	1
2.	400	40	4
3.	500	50	5
4.	60	6	0.6
5.	80	8	0.8

METRIC LINEAR UNITS			
	Centimeter	Decimeter	Meter
6.	200	20	2
7.	5	0.5	0.05
8.	150	15	1.5
9.	280	28	2.8
10.	90	9	0.9

Name ______________________________

Changing Units

You have to multiply or divide when you change units.

Multiply when you change from a larger unit to a smaller unit.

3 meters = ? dm
3 meters = ? cm

3 m × 10 = 30 dm
3 m × 100 = 300 cm

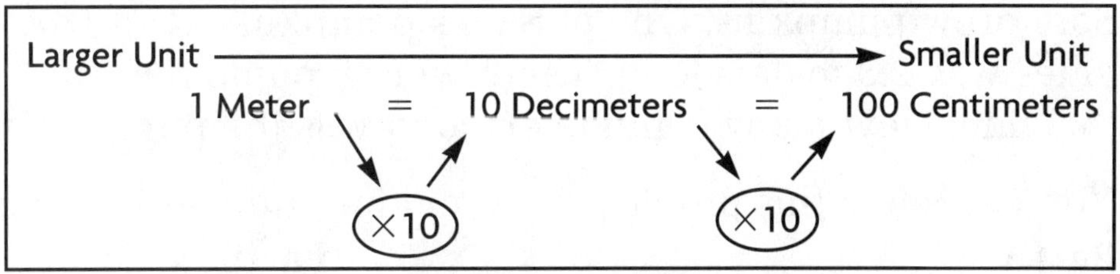

Divide when you change from a smaller unit to a larger unit.

300 cm = ? dm
300 cm = ? m

300 cm ÷ 10 = 30 dm
300 cm ÷ 100 = 3 m

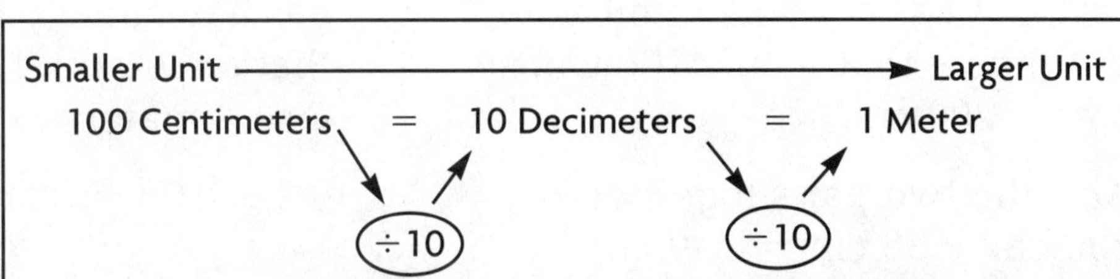

Circle the larger unit.

1. (meter) or decimeter

2. centimeter or (decimeter)

3. (meter) or centimeter

Would you multiply by 10 or by 100 to change the larger units to the smaller units? Write × *10* or × *100.*

4. 11 m = ? cm **× 100**

5. 20 dm = ? cm **× 10**

6. 8 m = ? dm **× 10**

Would you divide by 10 or by 100 to change the smaller units to the larger units? Write ÷ *10* or ÷ *100.*

7. 180 cm = ? dm **÷ 10**

8. 200 cm = ? m **÷ 100**

9. 800 dm = ? m **÷ 10**

Fill in the missing measures. You may use a calculator.

	Meters	Decimeters	Centimeters
10.	9	**90**	900
11.	**10**	100	1,000
12.	17	170	**1,700**
13.	**12**	120	1,200

Name ________________________________

LESSON 25.3

Problem-Solving Strategy

Solve a Simpler Problem

The Problem

Sam grew pumpkins. One of Sam's pumpkin vines was 2.3 meters long. There were 8 pumpkins on the vine. How many centimeters long was the pumpkin vine?

You can solve this problem by breaking it down into parts.

Part 1

Look at the whole-number part first. 2.3 has a whole number of 2. You know that 1 m = 100 cm, so 2 m = 200 cm.

Part 2

Now, look at the decimal part. 2.3 has a decimal part of 0.3. You know that a meter is 100 times larger than a centimeter, so multiply by 100.

$0.3 \times 100 = 30$ cm

Add the two parts together.
200 cm + 30 cm = 230 cm
Sam's pumpkin vine was 230 centimeters long.

Solve a simpler problem.

1. Lisa bought a jump rope that is 180 centimeters long. How many meters long is the jump rope?

   1.8 m long

2. Ms. Slats bought a new boat. It is 3.4 meters long. How many centimeters long is Ms. Slats's boat?

   340 cm

3. Larry's favorite baseball bat is 95 centimeters long. Moe's favorite baseball bat is 9.8 decimeters long. Whose bat is longer? How many centimeters longer?

   Moe's bat; 3 cm longer

4. The school janitor has to set up 320 chairs. He wants 16 or 32 chairs in a row. How many rows can he make?

   20 rows or 10 rows

5. In the past year, Mary's puppy grew 24 centimeters. Loretta's puppy grew 3.2 decimeters. Which puppy grew more in the past year? How many more centimeters?

   Lorretta's puppy; 8 cm more

6. Calvin and Sid both grow tomato plants. Calvin's plant is 11 decimeters tall. Sid's plant is 0.8 meters tall. Who has the taller plant? By how many decimeters?

   Calvin; 3 dm

Name ______________________________

Capacity

You can measure capacity, or how much liquid a container can hold, by using the units milliliter (mL), metric cup, or liter (L).

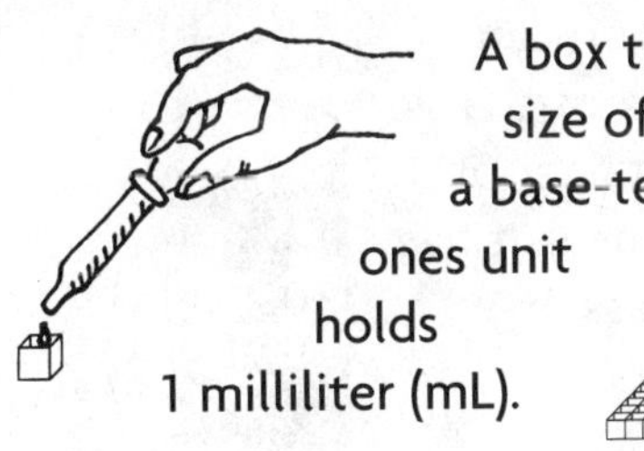

A box the size of a base-ten ones unit holds 1 milliliter (mL).

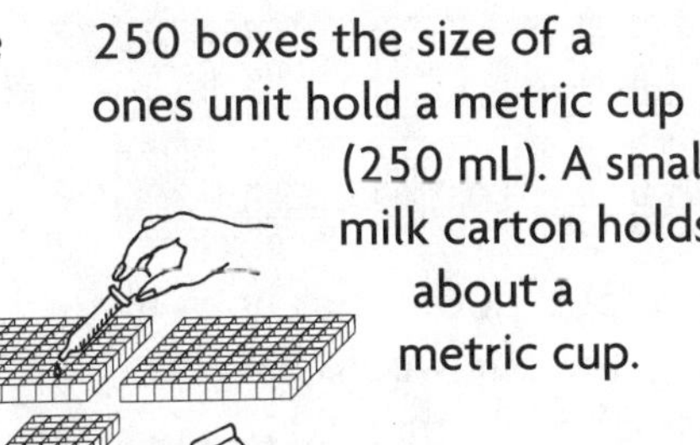

250 boxes the size of a ones unit hold a metric cup (250 mL). A small milk carton holds about a metric cup.

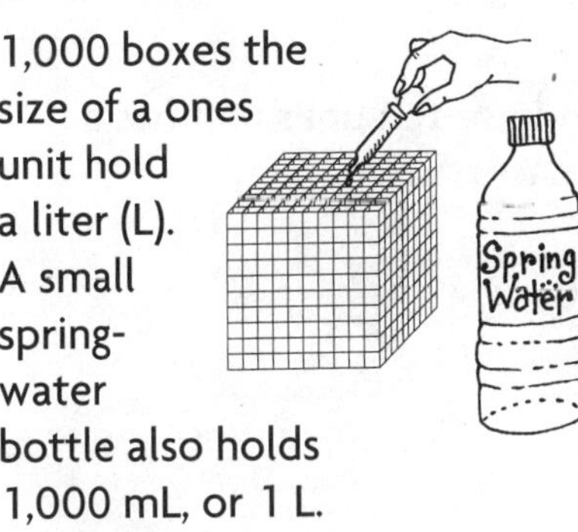

1,000 boxes the size of a ones unit hold a liter (L). A small spring-water bottle also holds 1,000 mL, or 1 L.

Use the references above to help you choose the most reasonable measure. Write a, b, or c.

1. 3 liters = [c] milliliters

   a. 30 mL b. 300 mL c. 3,000 mL

2. 500 milliliters = [a] metric cups

   a. 2 metric cups b. 20 metric cups c. 200 metric cups

3. 4,000 milliliters = [a] liters

   a. 4 L b. 40 L c. 400 L

4. 4 metric cups = [a] liter

   a. 1 L b. 10 L c. 100 L

5. 3 metric cups = [b] milliliters

   a. 75 mL b. 750 mL c. 7,500 mL

6. 2 liters = [c] milliliters

   a. 20 mL b. 200 mL c. 2,000 mL

Write the equivalent measurement.

7. 4 L = __4,000__ mL

8. 4 metric cups = __1,000__ mL

9. 8 L = __8,000__ mL

10. 8 metric cups = __2__ L

11. 500 mL = __2__ metric cups

12. 3 L = __12__ metric cups

Name ______________________________

Mass

You can measure mass with the units grams (g) and kilograms (kg).

A base-ten ones unit has a mass of 1 gram (g).

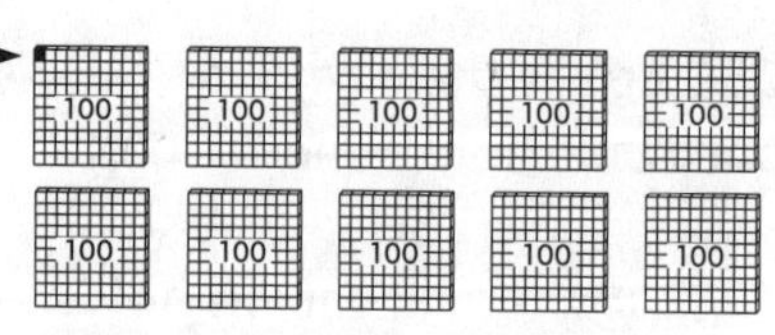

If you put 1,000 ones units in a bag, the bag would have a mass of 1 kilogram (kg), or 1,000 grams.

Or you can think of a large container of oatmeal. It has a mass of about 1 kg, or 1,000 g.

Circle each object that has a mass greater than 1 kg.

1. (a full-scale airplane)
2. a strand of hair
3. a pencil
4. (a bed)
5. (a full backpack)
6. a computer disk

Choose the more reasonable measurement. Write *g* or *kg*.

7. a shirt g
8. a photograph g
9. a refrigerator kg
10. a big bag of pet food kg
11. a baby brother kg
12. a pair of glasses g

Circle the more reasonable measurement.

13. spool of thread a. (10 g) b. 10 kg
14. car tire a. 20 g b. (20 kg)
15. bag of flour a. 5 g b. (5 kg)
16. hammer a. 2 g b. (2 kg)
17. scarf a. (25 g) b. 25 kg

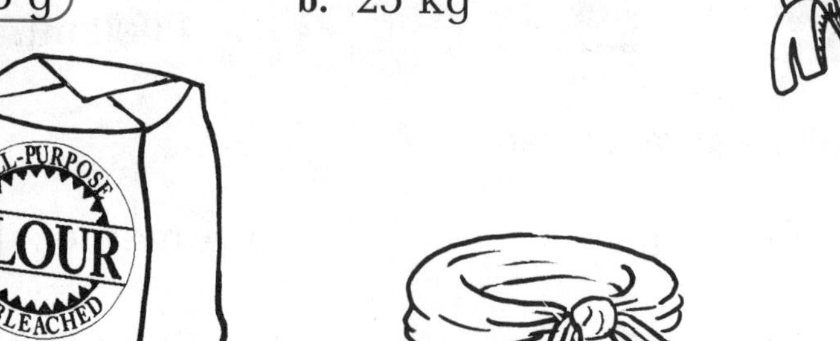

Name ____________________

Time as a Fraction

You can tell time using minutes or fractions.

30 minutes is one half of 60 minutes.

1:30 can also be read "half past one."

9:45 can also be read "a quarter to ten."

4:15 can also be read "a quarter past four."

15 minutes is one quarter of 60 minutes.

or

Draw the hands on the clock to show the time.

1.

3:30

2.

12:45

3.

4:15

4.

half past nine

5.

a quarter to five

6.

a quarter past seven

Write each time in a different way.

7. 10:30

half past ten

8. 6:45

a quarter to seven

9. 2:15

a quarter past two

10. half past six

6:30

11. a quarter to three

2:45

12. a quarter past eight

8:15

Name ____________________

Choosing Customary or Metric Units

You can measure the length of an object using different units.

An inch is a unit of measure in the customary system of measurement.

A centimeter is a unit of measure in the metric system of measurement.

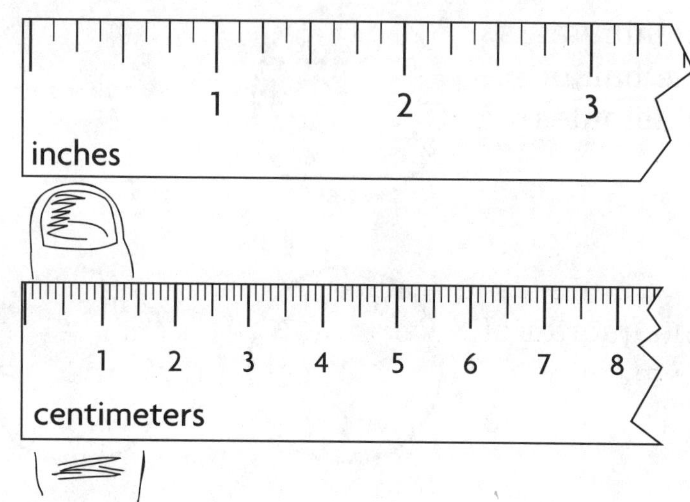

You can say the width of the finger is about $\frac{1}{2}$ inch or about $1\frac{1}{2}$ centimeters.

Measure each object in customary and metric units. **Measures may vary slightly.**

1. ____ 3 ____ in.

   ____ 7.5 ____ cm

2. ____ $1\frac{1}{4}$ ____ in.

   ____ 3 ____ cm

3. Width: ____ $\frac{3}{4}$ ____ in. ____ 1.9 ____ cm

   Height: ____ 1 ____ in. ____ 2.5 ____ cm

Circle the most reasonable measure for what is shown.

4. 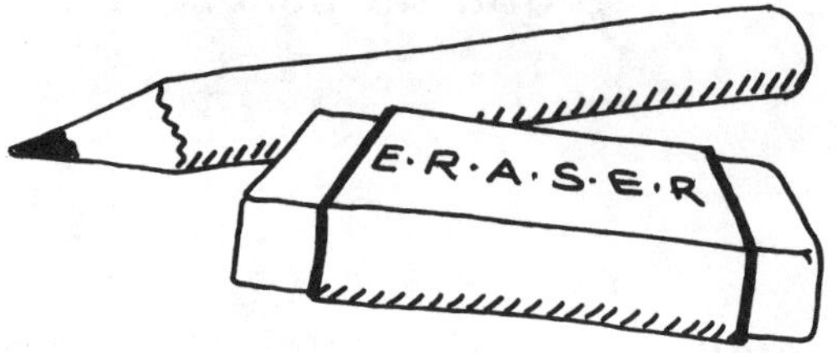

   Length of eraser

   (2 in.) or 2 cm

5. 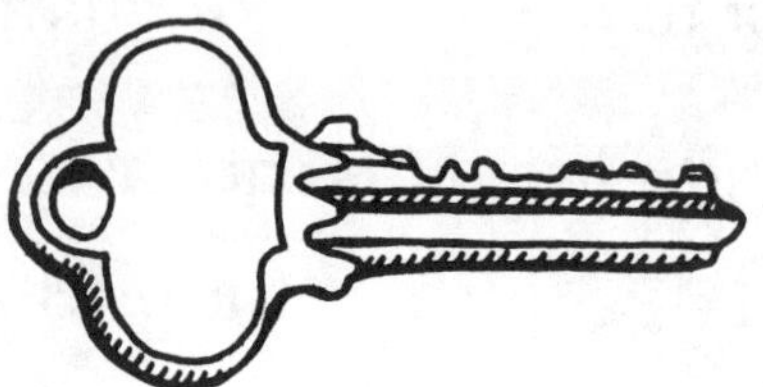

   Length of key

   (5 cm) or 5 in.

Name ______________________________

Problem-Solving Strategy

Write a Number Sentence

You can solve problems by writing a number sentence.

Pierre wants to build a fence around his pool. The fence needs to be 30 feet long and 18 feet wide. What is the perimeter of the fence in yards?

A number sentence can help you solve this problem. To find the perimeter in feet, add all the sides together.

$$30 + 30 + 18 + 18 = 96$$

A second number sentence changes the 96 feet to yards. There are 3 feet in a yard, so divide by 3.

$$96 \div 3 = 32$$

The perimeter of the fence is 32 yards.

Write two number sentences to solve each problem.

1. Patrice needs enough wall-paper border to complete his room. His room is square and measures 15 feet along each side. What is the perimeter of his room in yards?

   $15 \times 4 = 60$; 60 ft

   $60 \div 3 = 20$; 20 yd

2. Amy needs enough material for a border for a rug. The rug is 9 feet long and 6 feet wide. What is the perimeter of the rug in yards?

   $9 + 9 + 6 + 6 = 30$; 30 ft

   $30 \div 3 = 10$; 10 yd

3. The pool in Kathy's backyard is 5 yards long and 3 yards wide. What is the perimeter in feet?

   $5 + 5 + 3 + 3 = 16$; 16 yd

   $16 \times 3 = 48$; 48 ft

4. Allison bought 4 CDs priced at $12.00 each. The total cost, including tax, was $50.40. How much tax did Allison pay?

   \$12 × 4 = \$48

   \$50.40 − \$48.00 = \$2.40

Name ____________________

LESSON 26.3

Temperature

The customary unit for measuring temperature is **degrees Fahrenheit (°F)**.

The metric unit for measuring temperature is **degrees Celsius (°C)**.

Water freezes at 32°F, or 0°C. Water boils at 212°F, or 100°C. Room temperature is about 68°F, or 20°C.

The greater the temperature, the warmer it is. So, 80°F > 72°F.

Temperature can be compared by using subtraction. So, 80°F is 8° warmer than 72°F.

$$\begin{array}{r} 80^\circ \\ -\ 72^\circ \\ \hline 8^\circ \end{array}$$

Compare the temperatures. Write < or > to show which is warmer.

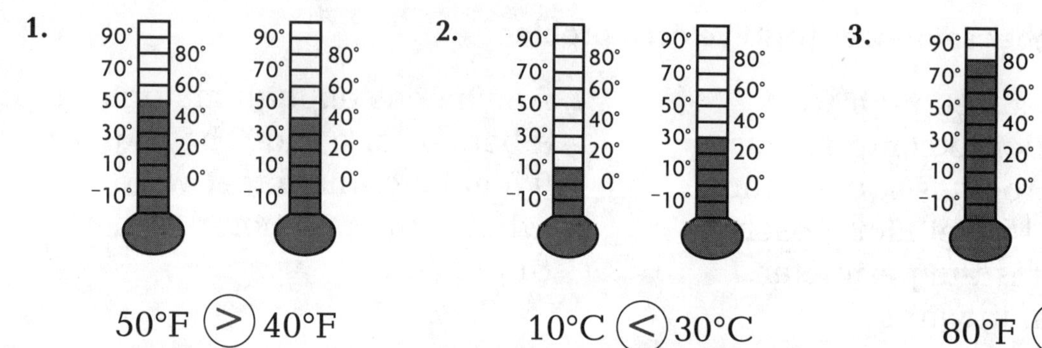

1. 50°F (>) 40°F
2. 10°C (<) 30°C
3. 80°F (>) 70°F

Write the temperatures. Write < or > to show which is warmer. Find the difference between the temperatures.

4. 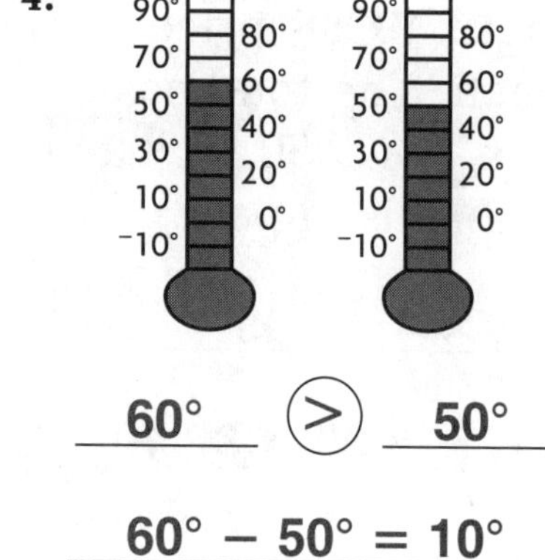

60° (>) 50°

60° − 50° = 10°

5. 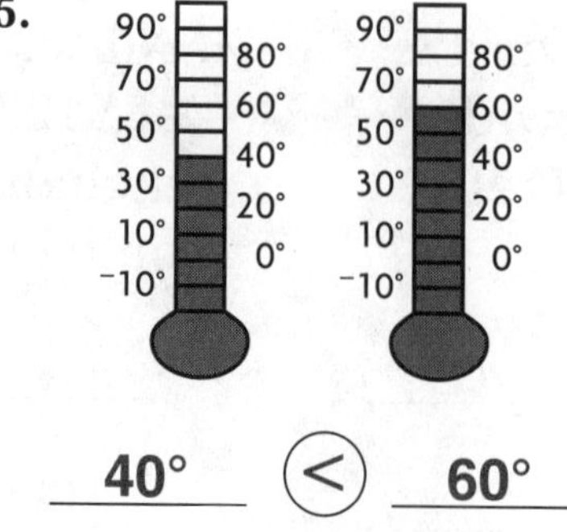

40° (<) 60°

60° − 40° = 20°

6. 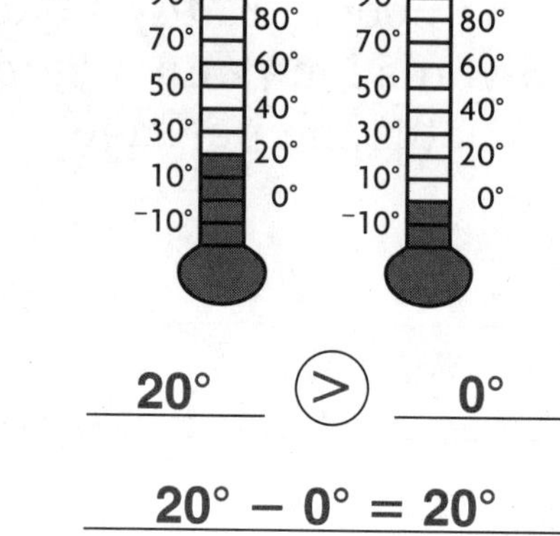

20° (>) 0°

20° − 0° = 20°

Name ______________________________

Making Equal Areas

You can combine shapes to make different shapes with the same area. For example,

+ =

+ =

You can also use shapes to model fractions. The triangle shows $\frac{1}{3}$ of the trapezoid.

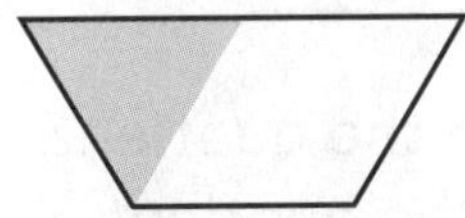

Trace and cut out the pattern-block shapes. Use them to model each area. **Possible answers are given.**

Pattern-Block Shapes

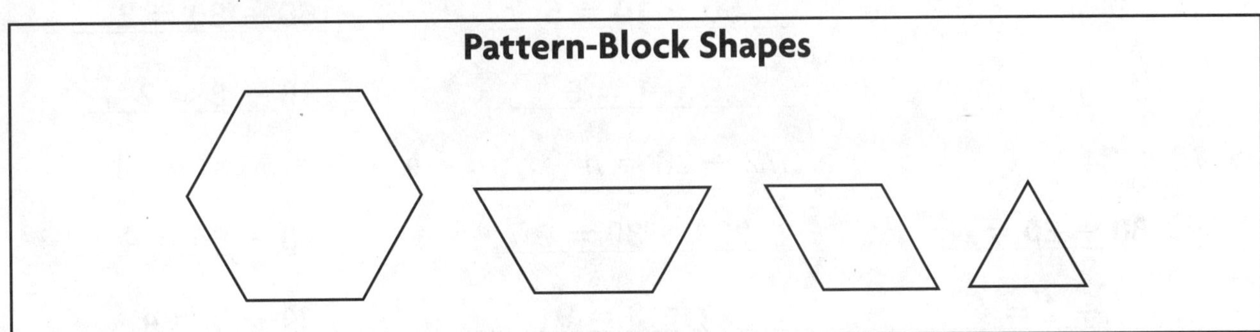

1. Model $\frac{1}{2}$ of ⬡.

2. Model $\frac{1}{2}$ of ▱.

3. Model $\frac{1}{3}$ of ⬡.

4. Model $\frac{1}{3}$ of ⏢.

5. Model $\frac{1}{6}$ of ⬡.

6. Model $\frac{2}{3}$ of ⬡.

Name ______________________

LESSON 27.1

Division Patterns to Estimate

In the problem $594 \div 18 = 33$, 594 is the dividend, 18 is the divisor, and 33 is the quotient. To estimate the quotient, follow Steps 1–3.

Step 1	Step 2	Step 3
Round the divisor to the nearest tens place.	Round the dividend to the nearest multiple of the new divisor.	Divide, using the rounded numbers.
$18 \rightarrow 20$	$595 \rightarrow 600$	$600 \div 20 = 30$

Write the numbers you would use to estimate the quotient. Then write the estimate. **Possible estimates are given.**

1. $36 \div 11 = n$ — **$40 \div 10 = 4$**
2. $78 \div 13 = n$ — **$80 \div 10 = 8$**
3. $56 \div 19 = n$ — **$60 \div 20 = 3$**

Write the numbers you would use to estimate the quotient. Then write the basic fact that helps you find the quotient. The first one has been done for you.

4. $446 \div 92 = n$
   $450 \div 90 = 5$
   $45 \div 9 = 5$
5. $52 \div 11 = n$
   $50 \div 10 = 5$
   $5 \div 1 = 5$
6. $162 \div 79 = n$
   $160 \div 80 = 2$
   $16 \div 8 = 2$
7. $76 \div 24 = n$
   $80 \div 20 = 4$
   $8 \div 2 = 4$
8. $272 \div 28 = n$
   $270 \div 30 = 9$
   $27 \div 3 = 9$
9. $235 \div 56 = n$
   $240 \div 60 = 4$
   $24 \div 6 = 4$

Complete the table.

	Dividend	Divisor	Quotient
10.	40	÷ 20	**2**
11.	**400**	÷ 20	20
12.	4,000	÷ 20	**200**

Name ______________________________

LESSON 27.2

Dividing by Tens

What is 387 ÷ 60? $60\overline{)3}$ with x above; $60\overline{)38}$ with xx above; $60\overline{)387}$ with xx? above

Think: What is the greatest multiple of 60 that can go into 3, 38, or 387?

Write the multiples of 60.

×	1	2	3	4	5	6	7
60	60	120	180	240	300	360	420

- All of the multiples of 60 are greater than 3 and 38.

  So, 60 × 6 = 360 is the greatest multiple of 60 that can go into 387.
- Divide. Place the number 6 in the quotient.

$$\begin{array}{r} xx6\ r27 \\ 60\overline{)387} \\ -360 \\ \hline 27 \end{array}$$

1. Complete the table to help you solve Exercises 2–5.

×	1	2	3	4	5	6	7	8	9
10	10	20	30	40	50	60	70	80	90
20	20	40	60	80	100	120	140	160	180
30	30	60	90	120	150	180	210	240	270
40	40	80	120	160	200	240	280	320	360
50	50	100	150	200	250	300	350	400	450
60	60	120	180	240	300	360	420	480	540
70	70	140	210	280	350	420	490	560	630
80	80	160	240	320	400	480	560	640	720
90	90	180	270	360	450	540	630	720	810

Find the quotient. Use the table to help you.

2. $30\overline{)233}$ **7 r23**
3. $50\overline{)419}$ **8 r19**
4. $20\overline{)189}$ **9 r9**
5. $40\overline{)353}$ **8 r33**

Name ______________________________

LESSON 27.3

Modeling Division

You can divide by using models.

What is 35 ÷ 12?

Step 1

Count out 35 objects.

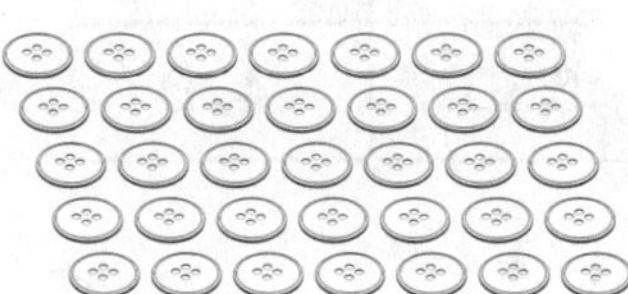

Step 2

Create as many groups of 12 as you can.

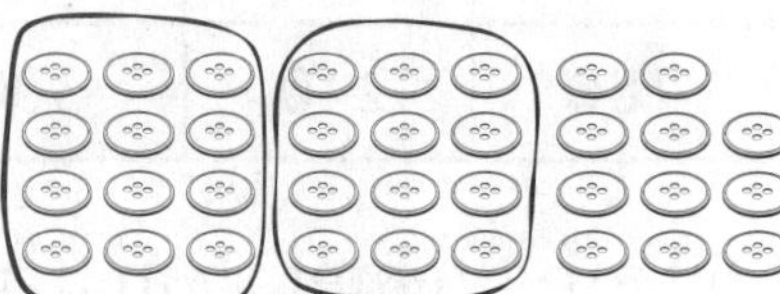

Step 3

Record your work. Count the number of groups and how many are left over.

$$\begin{array}{r} 2\text{ r}11 \\ 12\overline{)35} \\ -24 \\ \hline 11 \end{array}$$

Make a model and find the quotient. **Possible solutions are shown.**

1. $15\overline{)42}$ **2 r12**

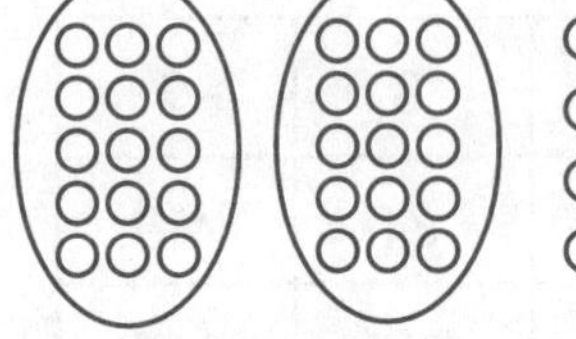

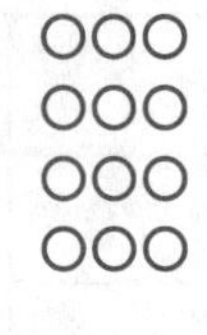

2. $17\overline{)49}$ **2 r15**

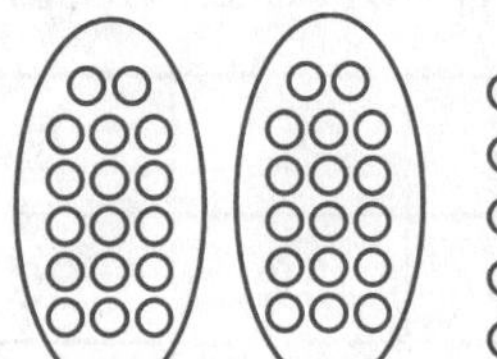

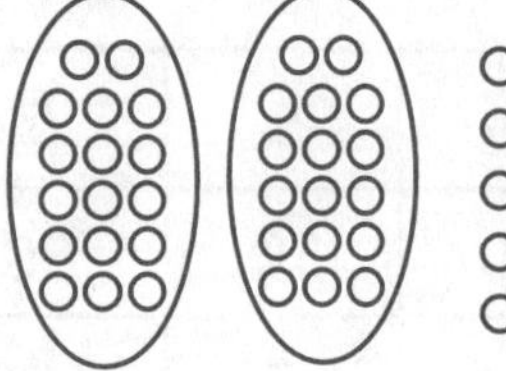

3. $25\overline{)103}$ **4 r3**

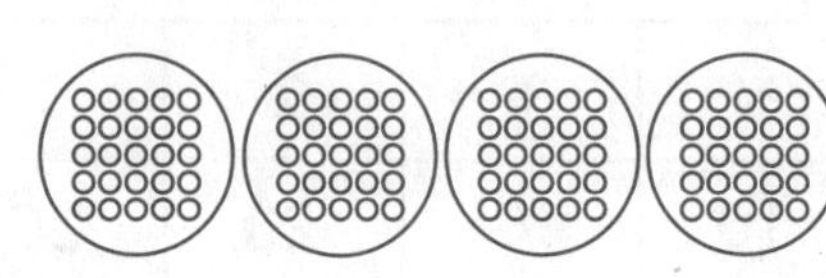

4. $21\overline{)65}$ **3 r2**

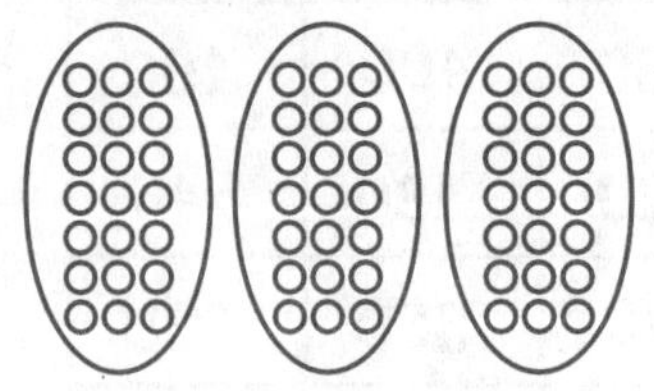

5. $11\overline{)35}$ **3 r2**

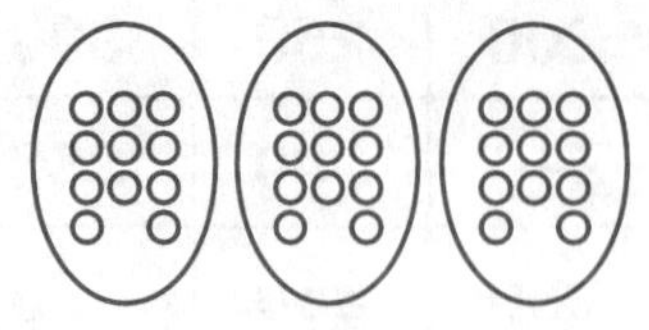

6. $31\overline{)128}$ **4 r4**

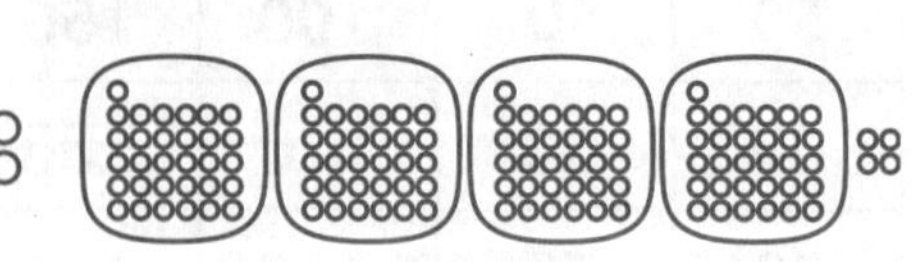

7. $16\overline{)93}$ **5 r13**

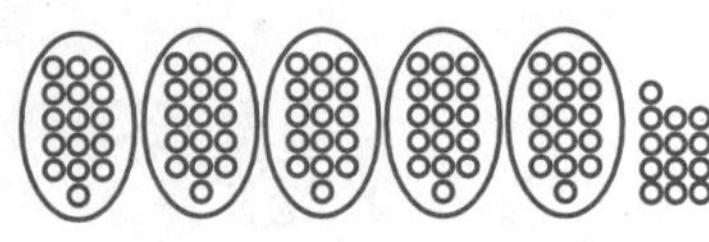

8. $19\overline{)59}$ **3 r2**

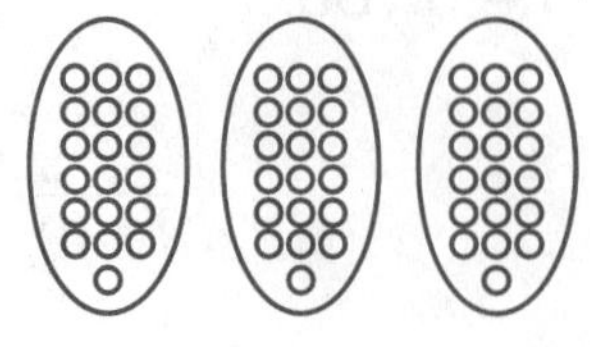

9. $24\overline{)84}$ **3 r12**

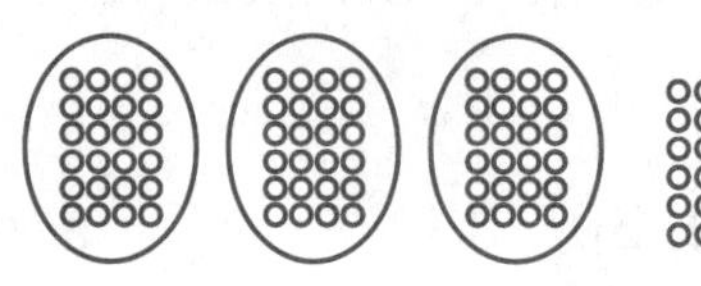

Name ________________________________

Division Procedures

What is 231 ÷ 19? Follow Steps 1–4 when you divide a 3-digit number by a 2-digit number.

Step 1

How many times will 19 go into 2? Zero times. Write an x in the hundreds place.

```
    x
19)231
```

Step 2

How many times will 19 go into 23? One time. Write a 1 in the tens place.

```
   x1
19)231
```

Step 3

Subtract 19 and bring down the 1.

```
   x1
19)231
  -19
   41
```

Step 4

How many times will 19 go into 41? Two times. Write a 2 in the ones place. Subtract.

```
  x12
19)231
  -19
   41
  -38
    3
```

Compare the remainder, 3, with the divisor, 19. The remainder is less than the divisor. Write the remainder as part of the quotient.

So, 231 ÷ 19 = 12 r3.

Find the quotient. Check by multiplying.

1. 11)806 — **73 r3**
2. 12)185 — **15 r5**
3. 15)122 — **8 r2**
4. 16)113 — **7 r1**
5. 17)414 — **24 r6**
6. 13)183 — **14 r1**
7. 18)581 — **32 r5**
8. 22)119 — **5 r9**

Name ____________________

Problem-Solving Strategy

Write a Number Sentence

Joan and Sam like to play catch with their dogs. Joan's dog can jump 6 feet into the air to catch a ball. Sam's dog can jump 68 inches into the air to catch a ball. How many feet can Sam's dog jump into the air? Which dog can jump higher into the air?

There are two questions to answer.

- To find how many feet Sam's dog can jump, divide the number of inches Sam's dog can jump by the number of inches in a foot.

  $$\begin{array}{r} 5\ \text{r}8 \\ 12\overline{)68} \\ -60 \\ \hline 8 \end{array}$$

  68 ÷ 12 = 5 r8 So, Sam's dog can jump 5 feet 8 inches.

- To find which dog can jump higher, compare the jumps, using the same units.

  Joan's dog can jump 6 feet. Sam's dog can jump 5 feet 8 inches. So, Joan's dog can jump higher.

Write a number sentence to solve. **Check students' work.**

1. What if Joan's dog jumps 7 feet and Sam's dog jumps 88 inches? What is the distance in feet that Sam's dog jumps? Which dog jumps higher?

   88 ÷ 12 = 7 r4;

   7 ft 4 in.; Sam's dog

2. Mark can blow up a balloon in 20 seconds. If he blows up balloons for 5 minutes, how many balloons can he blow up?

   5 × 60 = 300; 300 ÷ 20 = 15;

   15 balloons

3. Sid and Charles are brothers. Sid is 12 years 2 months old, and Charles is 118 months old. How many years old is Charles? Who is older, Sid or Charles?

   118 ÷ 12 = 9 r10;

   9 yr 10 mo; Sid

4. Tom solved the puzzle in 18 minutes. Shirley solved the puzzle in 950 seconds. How many minutes did it take Shirley to solve the puzzle? Who solved the puzzle faster?

   950 ÷ 60 = 15 r50;

   15 min 50 sec; Shirley

Name ________________________________

LESSON 27.5

Correcting Quotients

Divide. $321 \div 38 = n$ $\qquad 38\overline{)321}$

When you are dividing, sometimes the first digit in the quotient is too high or too low.

Too High	Too Low	Just Right
$\begin{array}{r} 9 \\ 38\overline{)321} \\ -342 \\ \hline \end{array}$	$\begin{array}{r} 7 \\ 38\overline{)321} \\ -266 \\ \hline 55 \end{array}$	$\begin{array}{r} 8 \\ 38\overline{)321} \\ -304 \\ \hline 17 \end{array}$
Since 342 > 321, the quotient 9 is too high. Use a lesser number.	Since the remainder, 55, is greater than the divisor, 38, the quotient 7 is too low. Use a greater number.	Since the remainder, 17, is less than the divisor, 38, the quotient 8 is just right.

Write *too high, too low,* or *just right* for each estimate. Explain.

1. $\begin{array}{r} 6 \\ 18\overline{)136} \\ -108 \\ \hline 28 \end{array}$ **Too low; the remainder is greater than the divisor.**

2. $\begin{array}{r} 7 \\ 16\overline{)103} \\ -112 \\ \hline \end{array}$ **Too high; the product is greater than the dividend.**

3. $\begin{array}{r} 8 \\ 17\overline{)139} \\ -136 \\ \hline 3 \end{array}$ **Just right; the remainder is less than the divisor.**

4. $\begin{array}{r} 5 \\ 23\overline{)119} \\ -115 \\ \hline 4 \end{array}$ **Just right; the remainder is less than the divisor.**

5. $\begin{array}{r} 6 \\ 32\overline{)236} \\ -192 \\ \hline 44 \end{array}$ **Too low; the remainder is greater than the divisor.**

6. $\begin{array}{r} 7 \\ 28\overline{)170} \\ -196 \\ \hline \end{array}$ **Too high; the product is greater than the dividend.**

Find the quotient.

7. $43\overline{)365}$ **8 r21**

8. $34\overline{)338}$ **9 r32**

9. $52\overline{)379}$ **7 r15**

10. $68\overline{)454}$ **6 r46**

Name ______________________________

LESSON 28.1

Making a Circle Graph

A **circle graph** is a graph in the shape of a circle. It shows data as a whole made up of different parts.

A fraction can be used to describe the size of each part.

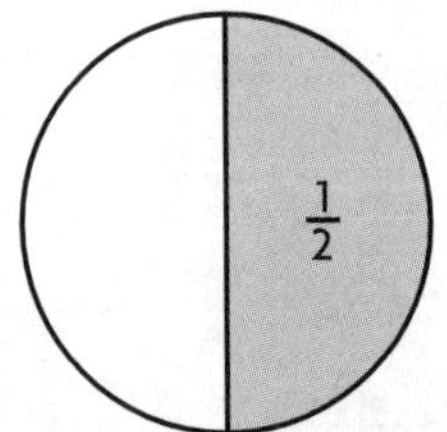

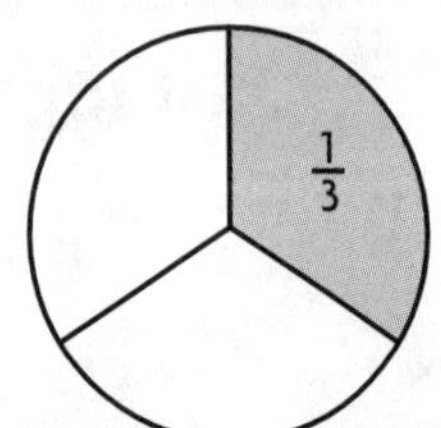

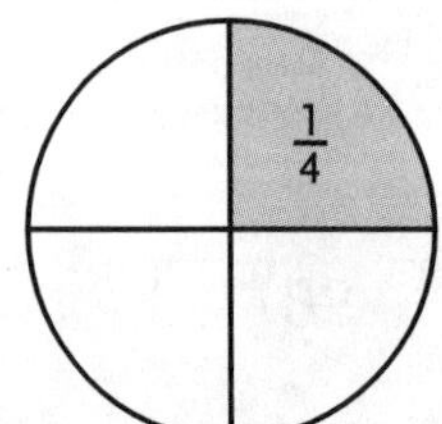

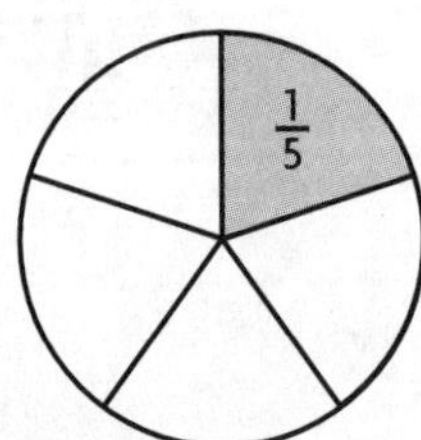

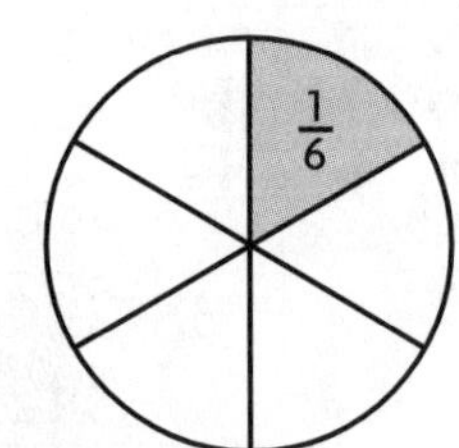

Write the fraction for the shaded part.

1.

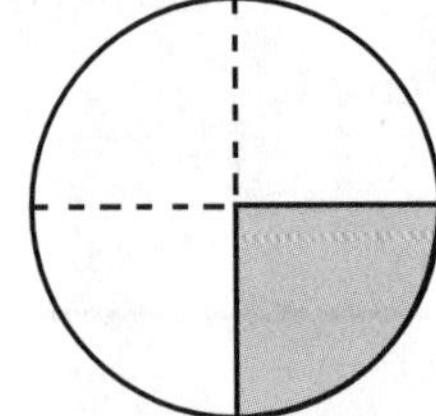

$\frac{1}{4}$

2.

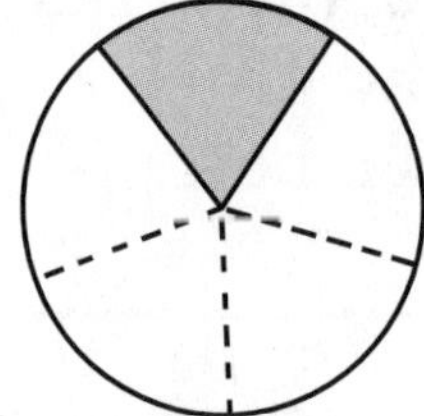

$\frac{1}{5}$

3.

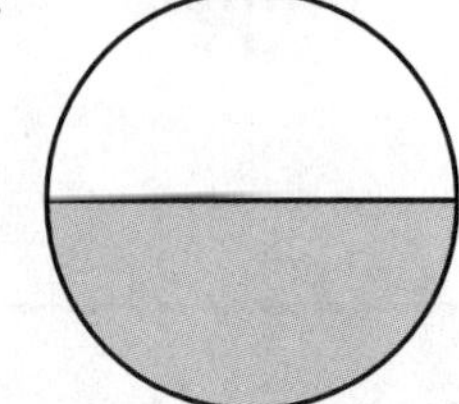

$\frac{1}{2}$

4.

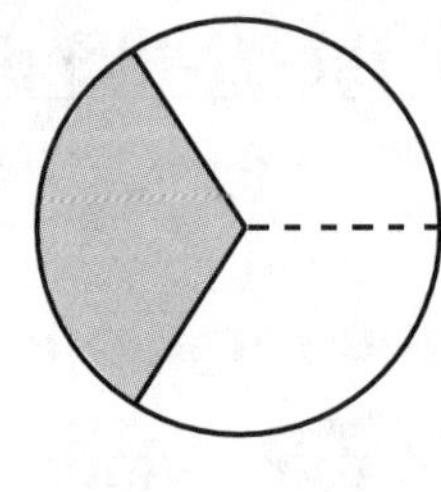

$\frac{1}{3}$

5.

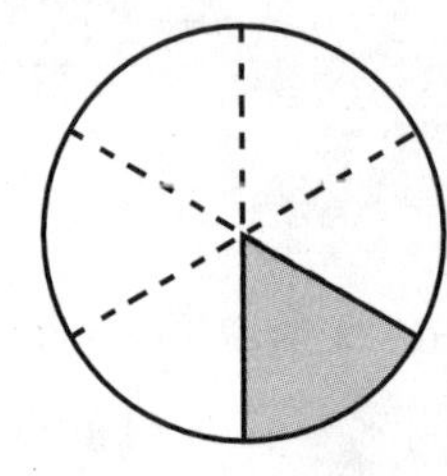

$\frac{1}{6}$

6.

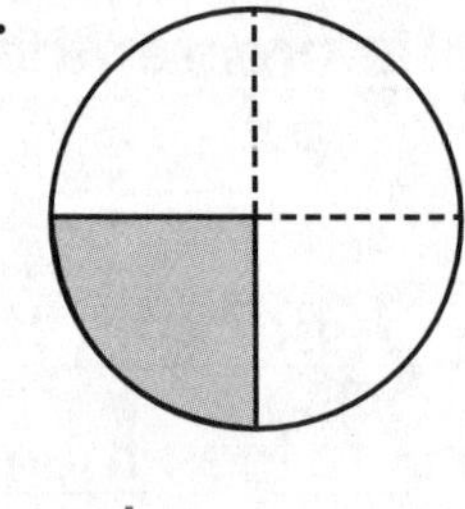

$\frac{1}{4}$

Label each graph.

7. a plate of cupcakes that is $\frac{1}{2}$ chocolate, $\frac{1}{3}$ vanilla, and $\frac{1}{6}$ chocolate chip

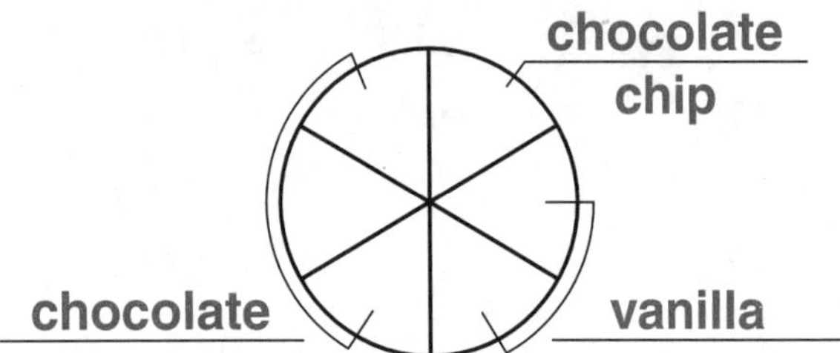

8. a state flag that is $\frac{1}{2}$ green, $\frac{1}{4}$ blue, and $\frac{1}{4}$ white

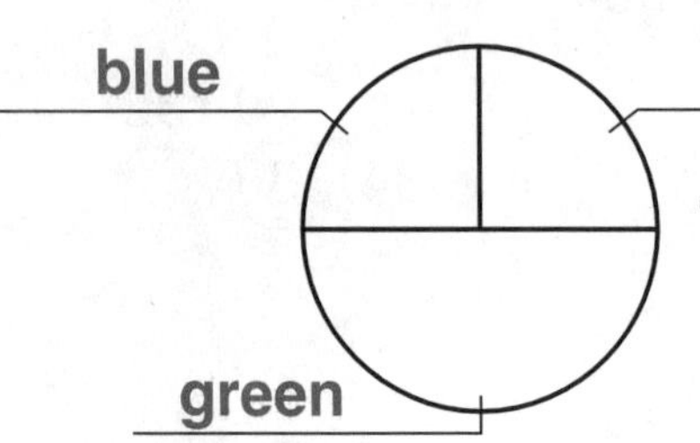

The placement of *white* and *blue* may vary.

Name ______________________

Fractions in Circle Graphs

Remember, when you write a fraction for the shaded part, the top number shows the number of parts shaded, and the bottom number shows the number of parts in all.

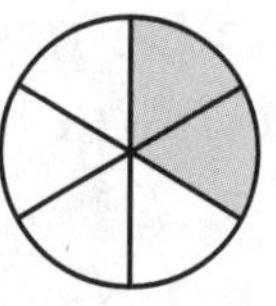

$\frac{2}{6}$ — 2: number of parts shaded; 6: number of parts in all

Some other examples:

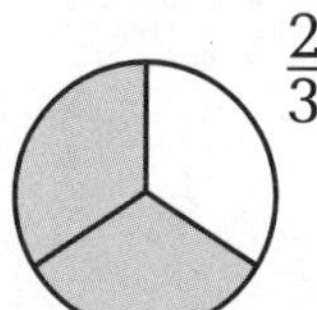
$\frac{2}{3}$

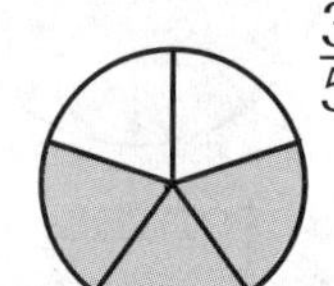
$\frac{3}{5}$

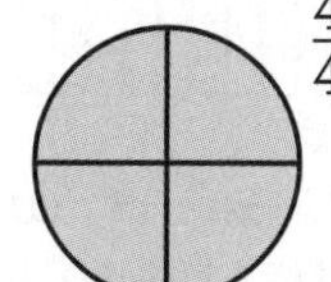
$\frac{4}{4}$

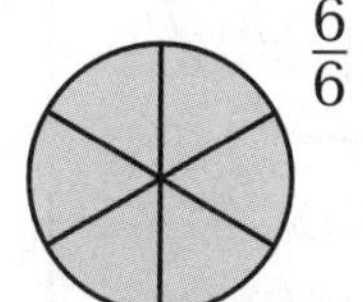
$\frac{6}{6}$

Write a fraction for the shaded part. Write a fraction for the unshaded part.

1.
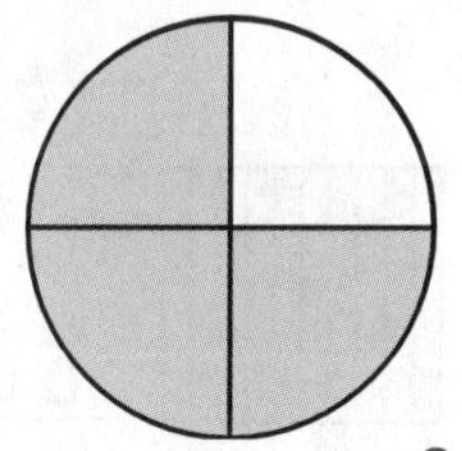
shaded $\frac{3}{4}$
unshaded $\frac{1}{4}$

2.
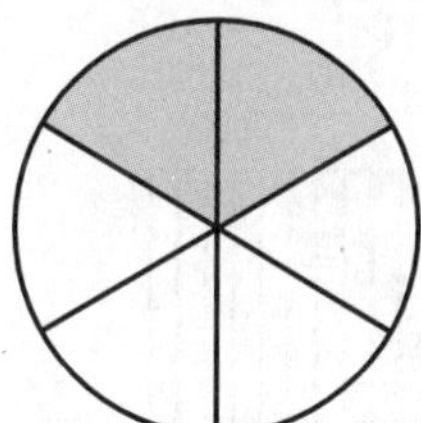
shaded $\frac{2}{6}$
unshaded $\frac{4}{6}$

3.
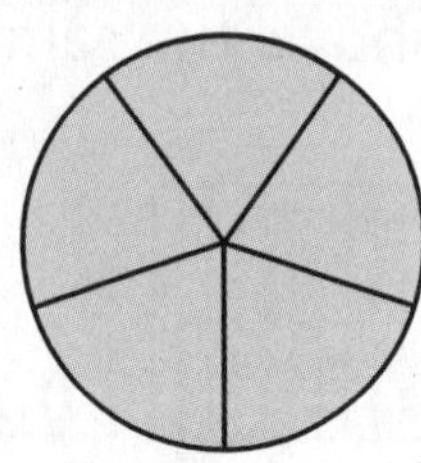
shaded $\frac{5}{5}$
unshaded $\frac{0}{5}$

4.
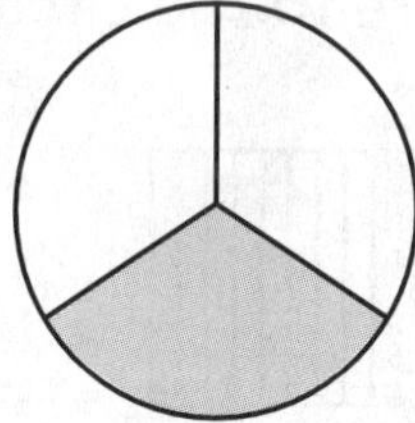
shaded $\frac{1}{3}$
unshaded $\frac{2}{3}$

Write a fraction for each part.

5.
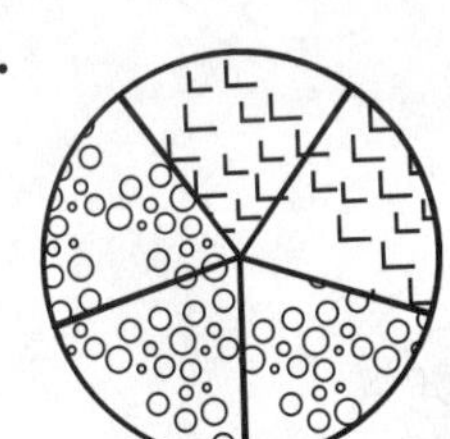
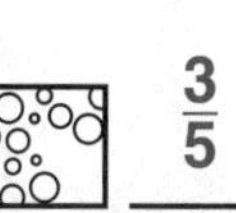
$\frac{2}{5}$
$\frac{3}{5}$

6.
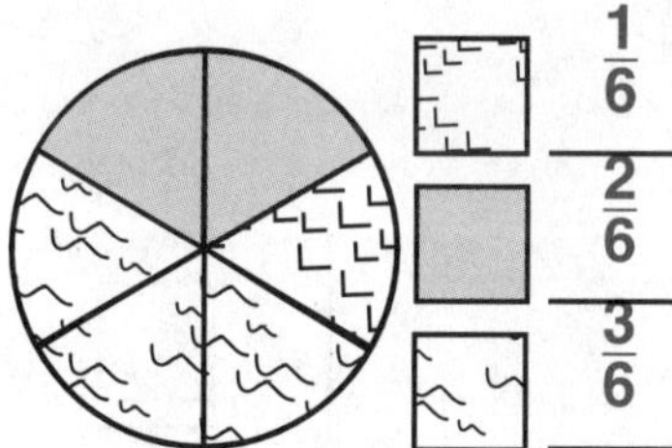
$\frac{1}{6}$
$\frac{2}{6}$
$\frac{3}{6}$

7.
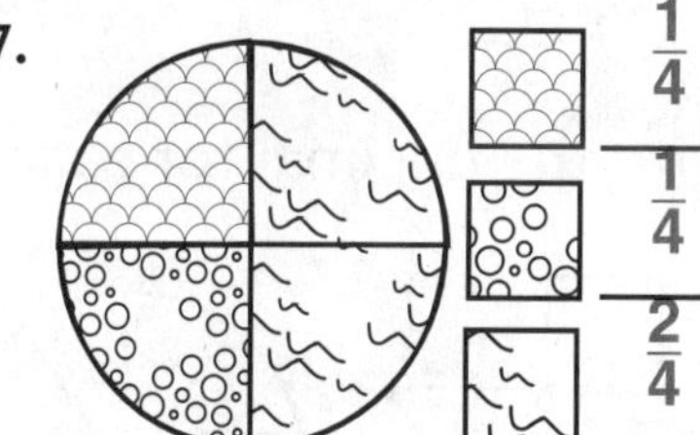
$\frac{1}{4}$
$\frac{1}{4}$
$\frac{2}{4}$

For Problems 8–10, use the circle graph.

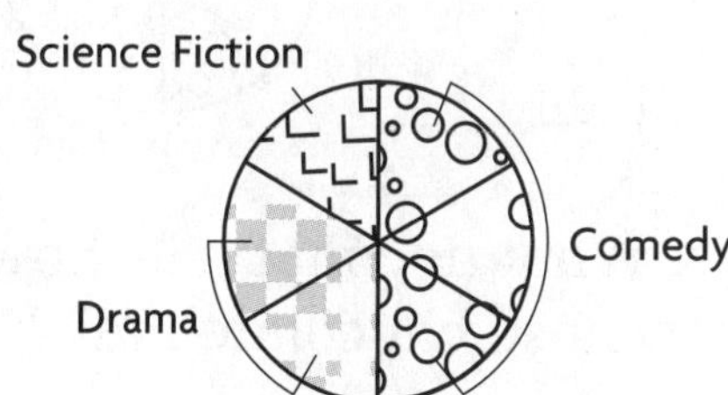

8. What fraction of the 30 movie videos are comedies? $\frac{3}{6}$, or $\frac{1}{2}$

9. Are more of the videos drama or science fiction? drama

10. What fraction represents all 30 videos? $\frac{6}{6}$

Name ______________________________

LESSON 28.3

Decimals in Circle Graphs

Just as decimals can be used to name parts of a square, they can be used to name parts of a circle.

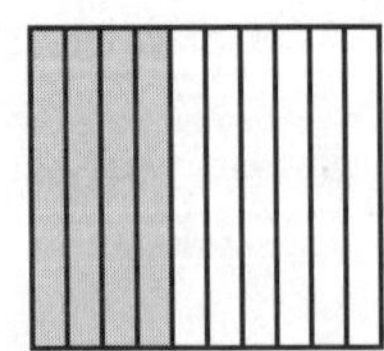

4 tenths, or 0.4, shaded

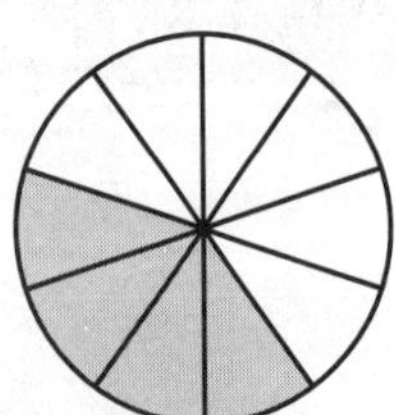

4 tenths, or 0.4, shaded

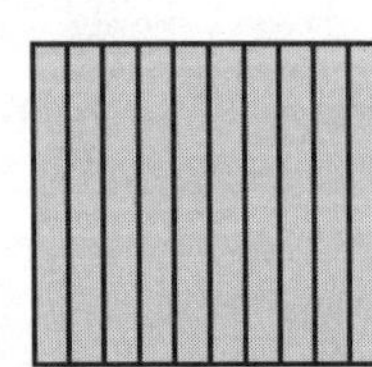

10 tenths, or 1.0, shaded

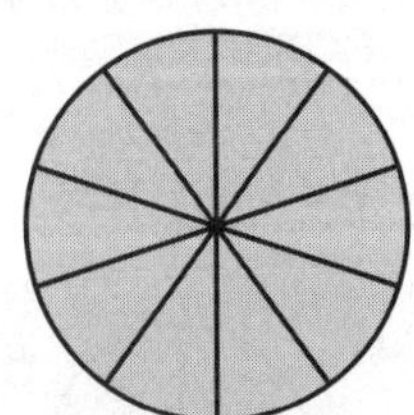

10 tenths, or 1.0, shaded

Write a decimal for the shaded part. Then draw a line from the square to the circle that shows the same decimal.

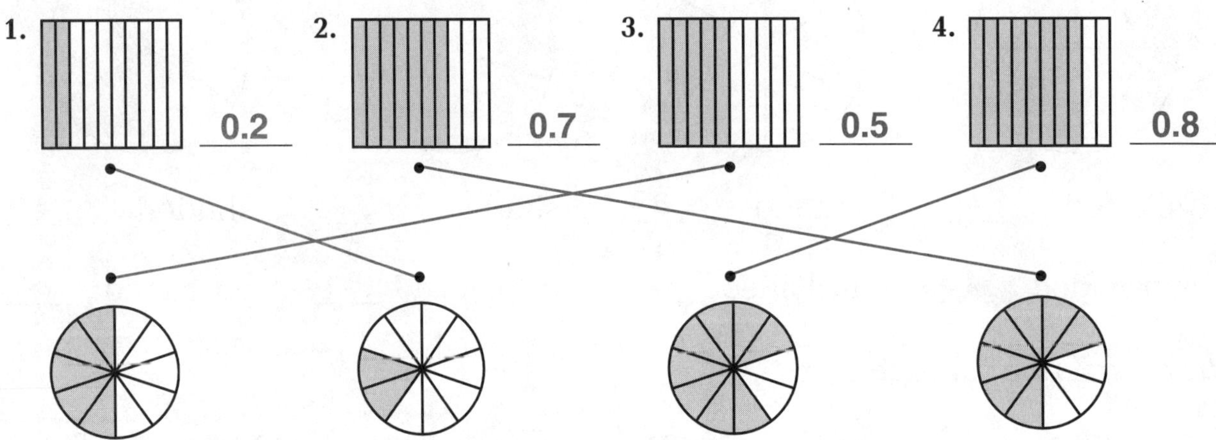

1. 0.2
2. 0.7
3. 0.5
4. 0.8

Write the decimal for each part.

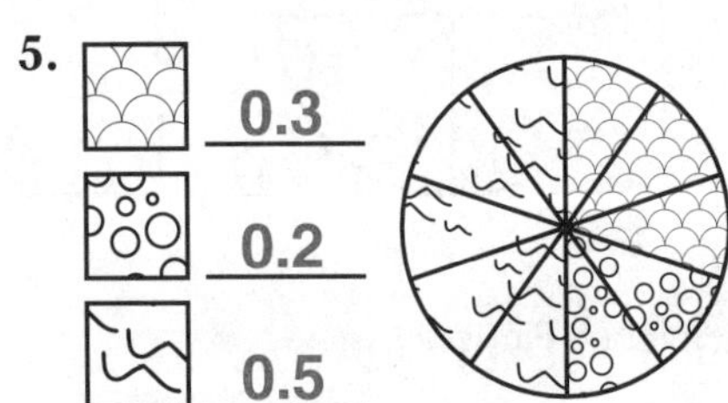

5. 0.3 — 0.2 — 0.5

6. 0.6 — 0.2 — 0.2

7. What decimal tells how many students chose chocolate milk?

0.6 students

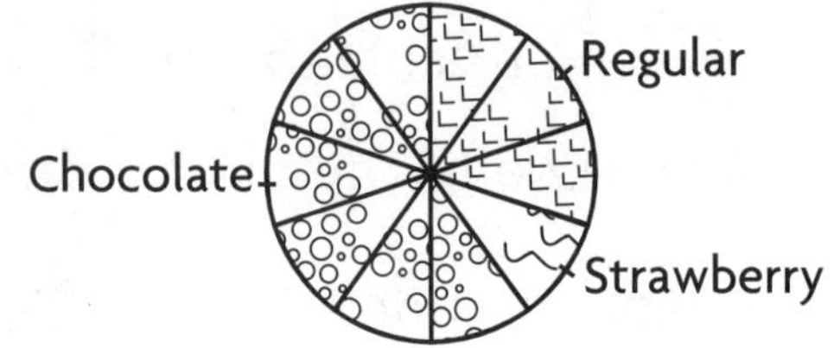

Favorite Milk Flavor of 10 Students

Name ____________________

Problem-Solving Strategy

Make a Graph

Two classes went on a field trip to Washington, D.C. Jo and Matt recorded the students' favorite places to visit in a chart, writing both fractions and decimals.

JO'S CLASS (10 STUDENTS)		
	Number of Students	
Place	Fraction	Decimal
White House	$\frac{1}{10}$	0.1
Washington Monument	$\frac{7}{10}$	0.7
Lincoln Memorial	$\frac{2}{10}$	0.2

MATT'S CLASS (10 STUDENTS)		
	Number of Students	
Place	Fraction	Decimal
White House	$\frac{3}{10}$	0.3
Washington Monument	$\frac{4}{10}$	0.4
Lincoln Memorial	$\frac{3}{10}$	0.3

To make a circle graph of the data, follow these steps.

Step 1 Write a title for the graph.

Step 2 For each group of data, color the graph to show the fraction or decimal. Use a different color for each group.

Step 3 Label each different-colored section of the graph.

Use the circles below and the information from the tables to make a circle graph for each class.

1. Lincoln Memorial — White House — Washington Monument

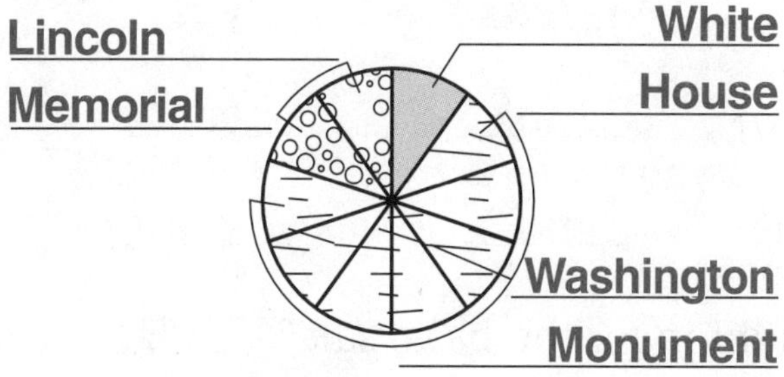

Choices of 10

Students in Jo's Class

2. Lincoln Memorial — White House — Washington Monument

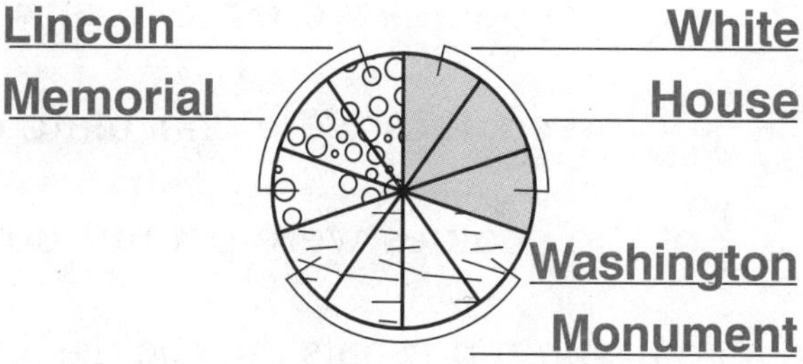

Choices of 10

Students in Matt's Class

Name ____________________

LESSON 28.4

Choosing Graphs to Represent Data

To choose the graph that represents a set of data, you must check to see that each part of the data is correctly represented in the graph.

Ms. Tucker's class voted on whether to watch a movie about polar bears, coyotes, or bald eagles. Of the class, $\frac{1}{2}$ voted for bald eagles, $\frac{1}{4}$ for polar bears, and $\frac{1}{4}$ for coyotes. Which graph represents how the class voted?

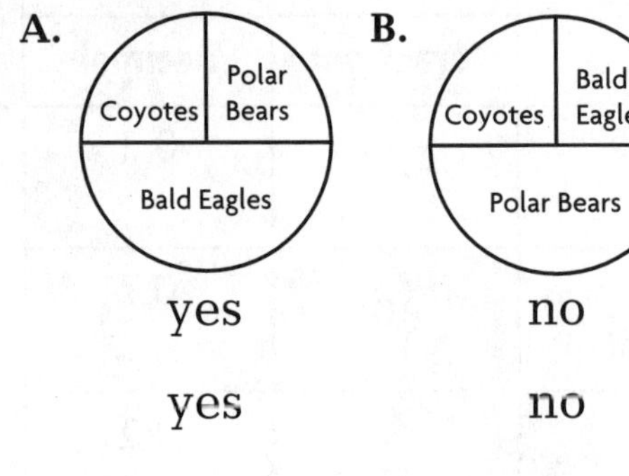

Does the graph show	A.	B.
$\frac{1}{2}$ of the votes for bald eagles?	yes	no
$\frac{1}{4}$ of the votes for polar bears?	yes	no
$\frac{1}{4}$ of the votes for coyotes?	yes	yes

So, Graph *A* shows how the class voted.

Read the description of the data, and then answer the questions.

Mrs. Mahajan's class brought in cookies for a bake sale. Of the cookies brought in, $\frac{1}{4}$ were molasses, $\frac{1}{2}$ were chocolate chip, and $\frac{1}{4}$ were peanut butter.

A.

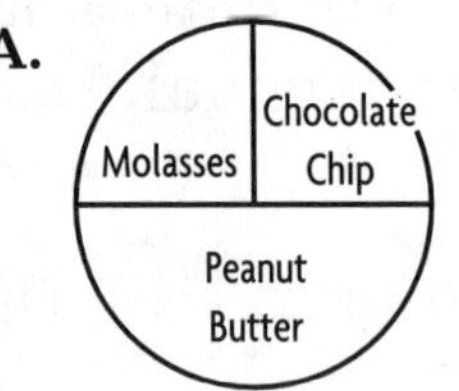

B.

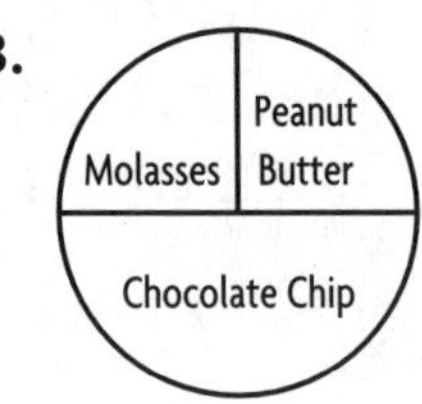

Does the graph show that

	A.	B.
1. $\frac{1}{4}$ of the cookies were molasses?	yes	yes
2. $\frac{1}{2}$ of the cookies were chocolate chip?	no	yes
3. $\frac{1}{4}$ of the cookies were peanut butter?	no	yes

Which graph represents the cookies brought in for the bake sale? B